EICOSANOIDS AND RADIATION

PROSTAGLANDINS, LEUKOTRIENES, AND CANCER

Series Editors: Kenneth V. Honn and Lawrence J. Marnett
Wayne State University School of Medicine
Detroit, Michigan

W.E.M. Lands, ed.: Biochemistry of Arachidonic Acid Metabolism. 1985. ISBN 0-89838-717-5.
L.J. Marnett, ed.: Arachidonic Acid Metabolism and Tumor Initiation. 1985. ISBN 0-89838-729-9.
S.M. Fischer and T.J. Slaga, eds.: Arachidonic Acid Metabolism and Tumor Promotion. 1985. ISBN 0-89838-724-8.
J.S. Goodwin, ed.: Prostaglandins and Immunity. 1985. ISBN 0-89838-723-X.
P. Polgar, ed.: Eicosanoids and Radiation. 1988. ISBN 0-89838-361-7.

EICOSANOIDS AND RADIATION

edited by

Peter Polgar
Boston University School of Medicine

Kluwer Academic Publishers
Boston/Dordrecht/London

Distributors for North America:
Kluwer Academic Publishers
101 Philip Drive
Assinippi Park
Norwell, Massachusetts 02061, USA

Distributors for the UK and Ireland:
Kluwer Academic Publishers
Falcon House, Queen Square
Lancaster LA1 1RN, UNITED KINGDOM

Distributors for all other countries:
Kluwer Academic Publishers Group
Distribution Centre
Post Office Box 322
3300 AH Dordrecht, THE NETHERLANDS

Library of Congress Cataloging-in-Publication Data

Eicosanoids and radiation / edited by Peter Polgar.
p. cm. — (Prostaglandins, leukotrienes, and cancer)
Includes bibliographies and index.
ISBN 0-89838-361-7
1. Prostaglandins—Effect of radiation on. 2. Arachidic acid—Derivatives—Effect of radiation on. 3. Ionizing radiation—Physiological effect. 4. Ultraviolet radiation—Physiological effect. I. Polgar, Peter. II. Series.
[DNLM: 1. Eicosanoic Acids—biosynthesis. 2. Eicosanoic Acids—metabolism. 3. Prostaglandins—radiation effects. QU 90 E34]
QP801.P68E37 1988
612′.01577—dc19
DNLM/DLC
for Library of Congress 87-31562
CIP

Printed in the United States of America

CONTENTS

CONTRIBUTORS

CATRAVAS, G.N., Department of Radiation Biochemistry and Chair of Science, Armed Forces Radiobiology Research Institute, Bethesda, Maryland 20814-5145, U.S.A.

DAS, U.N., Department of Clinical Pharmacology and Medical Research, Nizam's Institute of Medical Sciences, Punjagutta, Hyderabad 500482, India.

DELEO, V., Department of Dermatology, Columbia University, 630 West 168th Street, New York City, New York 10032, U.S.A.

GOLDBERG, I.D., Department of Radiotherapy, Harvard Medical School, 50 Binney Street, Boston, Massachusetts 02115, U.S.A.

KLIGMAN, L.H., Department of Dermatology, University of Pennsylvania, School of Medicine, Philadelphia, Pennsylvania 19104, U.S.A.

KOT, P.A., Department of Physiology and Biophysics, Georgetown University Medical Center, 3900 Reservoir Road N.W., Washington, D.C. 20007, U.S.A.

LIM, H.W., Department of Dermatology, New York University School of Medicine, New York City, New York 10016, U.S.A.

MENCONI, M., Boston University School of Medicine, Department of Biochemistry, Boston, Massachusetts 02118, U.S.A

MITCHELL, M.D., Department of Reproductive Medicine, University of California - San Diego, La Jolla, California 92093, U.S.A.

POLGAR, P., Boston University School of Medicine, Department of Biochemistry, Boston, Massachusetts 02118, U.S.A.

RAMA DEVI, G., Department of Clinical Pharmacology and Medical Research, Nizam's Institute of Medical Sciences, Punjagutta, Hyderabad 500482, Andhra Pradesh, India.

RAMWELL, P.W., Department of Physiology and Biophysics, Georgetown University Medical Center, 3900 Reservoir Road N.W., Washington, D.C. 20007, U.S.A.

REICHEL, R., First Department of Gynecology and Obstetrics, University of Vienna, Spitalgasse 23, A-1097 Vienna, Austria.

ROSEN, E.M., Department of Therapeutic Radiology, Yale University Medical School, 333 Cedar Street, New Haven, Connecticut 06510, U.S.A.

RUBIN, D.B., Rush University Medical School, Department of Medicine, Rush Presbyterian St. Luke's Medical Center, Chicago, Illinois 60612 U.S.A.

SCHNEIDKRAUT, M.J., Department of Physiology and Biophysics, Georgetown University Medical Center, 3900 Reservoir Road N.W., Washington, D.C. 20007, U.S.A.

SINZINGER, H., Department of Nuclear Medicine and Atherosclerosis Research Group (ASF), University of Vienna, Schwarzspanierstrabe 17, A-1090 Vienna, Austria.

STEEL, L.K., Department of Radiation Biochemistry and Chair of Science, Armed Forces Radiobiology Research Institute, Bethesda, Maryland 20814-5145, U.S.A.

STRICKLAND, D.M., Division of Maternal and Child Health, Wilford Hall USAF Medical Center, San Antonio, Texas 78236, U.S.A.

VINTER, D.W., Department of Radiotherapy, Harvard Medical School, 50 Binney Street, Boston, Massachusetts 02115, U.S.A.

WALDMAN, J.S., Dermatology Section, N.Y. Veterans Administration Medical Center, New York City, New York 10016, U.S.A.

WEPPLEMAN, B., Department of Radiation - Oncology, University of Alabama at Birmingham, 619 South 19th Street, Birmingham, Alabama 35233, U.S.A.

FOREWORD

Prostaglandins, Leukotrienes, and Cancer is a multi-volume series that will focus on an emerging area of cancer research. In 1968, R.H. Williams first reported that elevated prostaglandin levels are present in human medullary carcinoma. Since that time, the concept that arachidonic acid metabolites may be involved in cancer has expanded to include every aspect of the disease from cell transformation through metastasis.

Prostaglandins and leukotrienes are generic terms used to describe a family of bioactive lipids produced from unsaturated fatty acids (principally from arachidonic acid) via the cyclooxygenase and lipoxygenase pathways, respectively. Cyclooxygenase products consist of diverse products such as prostaglandin E_2 (PGE_2), prostacyclin (PGI_2) and thromboxane A_2 (TXA_2), whereas lipoxygenase products consist of hydroperoxy fatty acids and mono-, di- and tri-hydroxy acids including leukotrienes. The precursor fatty acids for the cyclooxygenase and lipoxygenase pathways are present in cellular phospholipids. This finding established an important control point in their biosynthesis—the release of substrate. This occurs in response to numerous stimuli that act at the cell surface. Dr. Bengt Samuelsson's extensive study of the metabolism of prostaglandins indicated that they are rapidly inactivated on a single pass through pulmonary circulation. Thus, they cannot act as circulating hormones and appear to be made on demand in or in the vicinity of target tissues leading to the concept that prostaglandins are local hormones or autocoids.

Altered production, qualitative and/or quantitative, of prostaglandins and leukotrienes has been implicated in the development of a number of disease states (e.g., atherosclerosis, inflammatory diseases, asthma). Evidence has been accumulating in the literature suggesting that prostaglandins and leukotrienes may stimulate or inhibit various steps in the complex etiology of cancer, i.e., steps in the progression from a transformed cell to a metastatic tumor. The initial volumes in this series will examine the roles of prostaglandins and leukotrienes in tumor initiation, tumor promotion, tumor cell growth and differentiation, tumor immunity, tumor metastasis and cancer therapy. We hope as this field of cancer research develops that this series, Prostaglandins, Leukotrienes, and Cancer, will provide a forum within the framework of current evidence for the synthesis of new hypotheses and discussion of controversial issues.

Kenneth V. Honn
Lawrence J. Marnett

PREFACE

Convincing evidence has now accumulated linking radiation and eicosanoid production. The reported radiation caused changes have varied from those showing overall increases in prostaglandin (PG) production, to shifts in the production of PG types, to general decreases in PG production. Because of the complexity of dosimetry, time of measurement after exposure to radiation and the given tissue or cell type, many of these apparent conflicting data may in reality represent differences in experimental conditions. Clearly, further experimental data is necessary to fully understand the process. What is important at this time is to recognize the fact that both ionizing and ultraviolet radiation may be affecting the synthesis of eicosanoids in a physiologically significant manner. Interestingly, data now exists suggesting that the presence of eicosanoids offers the organism protection from radiational injury. In light of these reports, it is conceivable that a better understanding of the interaction between eicosanoids and radiation will ultimately prove to be therapeutically beneficial.

In this volume, we present a comprehensive discussion of eicosanoids and radiation. We concentrate on ionizing and ultraviolet radiation, if for no other reason than that much of the existing data has focused on these two areas. As much as possible, we attempt to present data as it has developed, including conflicting results. The effect of ionizing radiation on eicosanoid metabolism is discussed in four chapters in terms of cells in culture, intact tissue, the whole organism and blood vessels. A similar approach is used to examine the effects of ultraviolet radiation on eicosanoid metabolism.

Evidence showing that eicosanoids afford the organism protection against both ionizing and ultraviolet radiation is discussed in a series of articles, including evidence that eicosanoids provide protection against genetic damage.

Radiation is generally viewed from the perspective of nucleus and chromosomal alterations. However, lipid peroxidation and the generation of reactive radicals may also be important factors in the radiational effects on cell physiology. In fact, the generation of peroxide molecules and oxygen radicals are important elements in the regulation of a number of enzyme systems, including those associated with eicosanoid synthesis. This relationship is discussed in a separate chapter where we examine the influence of peroxides and oxygen radicals on both the cyclooxygenase and phospholipase A_2 enzyme systems. Finally, to place the eicosanoid story into perspective, the last two chapters discuss the radiobiology of blood vessels and the effect of ultraviolet radiation on the skin.

It is the aim of this volume to summarize and clarify the interactions between radiation and eicosanoids as they are presently understood.

EICOSANOIDS AND RADIATION

1

PROSTAGLANDINS AND RADIATION IN CELL CULTURE

David B. Rubin, M.D.
Rush University Medical School
Chicago, IL

The advantage of the cell culture approach over in vivo methods is that one is able to examine physiological mechanisms in isolated cell types by designing experiments with fewer confounding variables. As one might expect, the cell biology of prostaglandin production and treatment has been extensively characterized in cultured cells. The cell culture approach also has been invaluable in our understanding of radiation biology, but the literature on the interaction of radiation and eicosanoids in cultured cells surprisingly is not extensive. A brief review of this topic is included in a chapter by Polgar et al. (1). Production of two prostaglandins have been measured in irradiated cells. PGI_2 by radioimmunoassay for 6-keto-PGF_{α}, and total PGE by radioimmunoassay for PGE_1 and PGE_2. The goals of these studies have been to determine the role of eicosanoids 1) as markers of cytotoxicity, 2) in relationship to acute cell viability and long-term proliferative survival, 3) in relationship to receptors and cell/cell interactions, and 4) as radioprotective agents.

PROSTAGLANDINS AS MARKERS OF RADIATION-INDUCED CYTOTOXICITY

After irradiation, there are two ways to examine de novo synthesis of prostaglandins in cell cultures: net accumulation in medium over extended periods (hours to days), and incremental capacity to produce prostaglandins for a short interval (minutes). Measurements of incremental capacity of prostaglandin production are made in cultures whose medium has been changed, cells rinsed, and fresh medium added for an increment of time after irradiation. Net accumulation of prostaglandins in culture medium is usually expressed as picograms or nanograms per ml. It is not clear by this approach how much prostaglandin is produced by membranes of living cells as compared to destroyed cells. This point is particularly important for mammalian cells (such as those from the vascular endothelium) that are viable when they are

P. Polgar (ed.), EICOSANOIDS AND RADIATION.

adherent and generally not viable when detached. However, the incremental capacity to produce prostaglandin over 2 to 10 minutes can be corrected for the number of adherent (viable) cells, since a significant amount of cell loss does not occur over the short interval. This assumption allows one to express incremental prostaglandin production on the basis of cell number or contents of DNA or protein. The incremental capacity of prostaglandin production can be examined with and without the stimulation of synthesis, and can be related to the metabolic function of viable cells. Measurement of incremental capacity may be more relevant than net accumulation of prostaglandins in understanding their role in the cytotoxic response of tissue in vivo.

The net accumulation of prostaglandins in medium of irradiated cultures has been closely linked to cell destruction. PGE increases in the medium of irradiated marrow stromal cells 3 days after the cells received 100-200 Gy of gamma rays (2) and after 4-24 hours, 5-50 Gy increased PGI_2 accumulation in medium of irradiated endothelial cells cultured from bovine aorta (3,4). In these three studies, the amount of cell lysis was directly related to the level of prostaglandin per ml. In one study (4), reduction of cell number protein and DNA contents, as well as release of lactate dehydrogenase activity into the medium (a marker of cytoplasmic leakage) changed in parallel to PGI_2 accumulation in the medium (Figure 1). Gibson et al. also have reported (2) that net accumulation of PGE over 3 days in cultures of marrow stromal cells is lowered in a dose-related fashion (1-8 Gy) but this observation was reported on replicate samples from one experiment and was not verified in a repeated experiment.

Radiation has been reported to both increase (4-7) and decrease (3,4) the incremental PGI_2 production in cultured vascular endothelial cells. The change in prostaglandin production is likely related to the time after irradiation. Hahn and colleagues first had published reports (5,6) on an increase in the capacity for PGI_2 production, whereas Eldor and coworkers later reported (3) about a decrease . This decrease was related to radiation dose (3-12 Gy) and time (over 21 days) and fractionated doses were less inhibitory than a single dose. Both groups used similar doses of irradiation (below 10 Gy) but Hahn et al.examined cultures 3 to 7 hours post irradiation and Eldor et al. made their observations after 24 hours. My coworkers and I (4) found both observations to be true (Figure 2): an increase PGI_2 production 4 to 24 hours after 5 Gy and a reduced production at 48 and 72 hours. Furthermore, arachidonate (4) and bradykinin (5,6) enhanced the

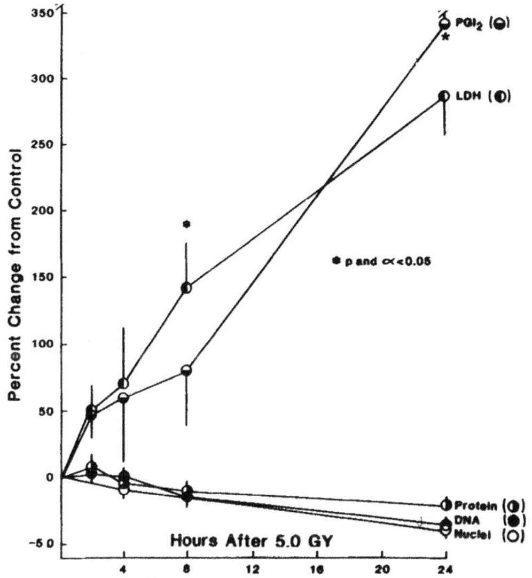

FIG. 1. Nuclei count (o), and deoxyribonucleic acid (DNA) (●) and protein (◐) concentration in monolayer; lactate dehydrogenase (LDH) activity (◐) and prostacyclin (PGI_2) concentration (◐) in culture medium (2% serum) of bovine aortic endothelial cells as a function of time after exposure to 5.0 Gy 60 Co γ-rays. PGI_2 data reflect net accumulation of 6-ketoprostaglandin F_1 (6-keto-$PGF_{1\alpha}$) without arachidonate supplementation. Data are expressed relative to comparably aged sham-irradiated cells and represent means $\pm$ SE of 8 observations per group. *Significant changes for all end points (Reprinted with permission, J. Appl. Physiol. 58, 592-597, 1985).

radiation-induced augmentation in PGI_2 production.

The biphasic change in incremental capacity of prostanoid production can be explained by a number of mechanisms. The initial increase by intact cells could be caused by the oxidant stress created by irradiation. According to Hemler et al. (8), prostaglandin production can be driven by the "peroxide tone" of a membrane which should be increased following irradiation. Hahn and coworkers (6) and Friedman et al. (7) argue that the mechanism for enhanced PGI_2 production relates to the increased release and availability of arachidonate. Hahn et al. suspect that, as a result of irradiation, arachidonate becomes more available for prostaglandin production, that there is no significant inhibition of arachidonate reesterfication and that cyclo-oxygenase activity is increased. The increase in cyclooxygenase activity did not appear to be related to the de novo synthesis of the enzyme, as inhibition of protein synthesis by cyclohexamide did not lower the augmentation of PGI_2 production in irradiated cells treated with bradykinin.

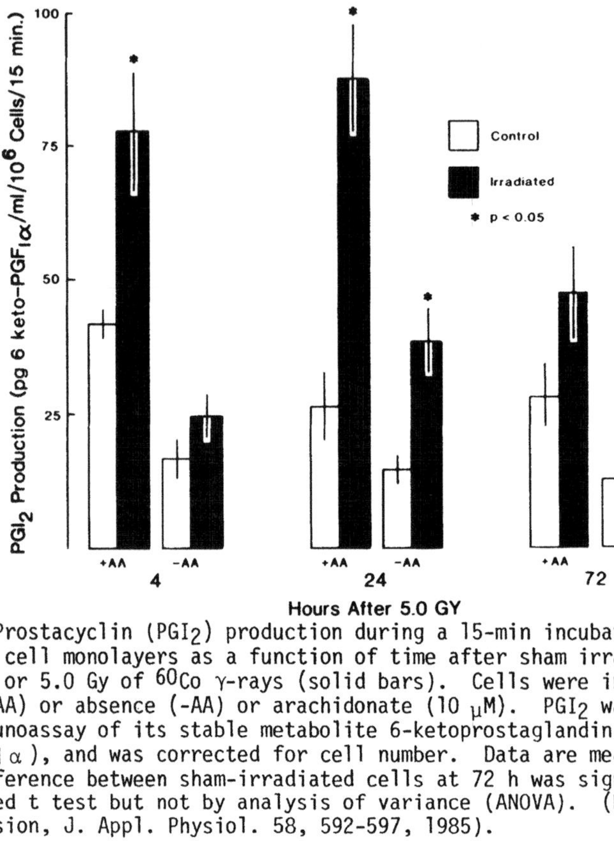

FIG. 2. Prostacyclin (PGI_2) production during a 15-min incubation of endothelial cell monolayers as a function of time after sham irradiation (open bars) or 5.0 Gy of ^{60}Co γ-rays (solid bars). Cells were incubated in presence (+AA) or absence (-AA) or arachidonate (10 μM). PGI_2 was measured by radioimmunoassay of its stable metabolite 6-ketoprostaglandin $F1_{\alpha}$ (6-keto-$PGF1_{\alpha}$), and was corrected for cell number. Data are means $\pm$ SE; n = 6. Difference between sham-irradiated cells at 72 h was significant by the paired t test but not by analysis of variance (ANOVA). (Reprinted with permission, J. Appl. Physiol. 58, 592-597, 1985).

According to Eldor et al. (3), the reduction in PGI_2 production over time can be caused by damage to the enzymatic mechanism required for prostaglandin production. As evidence for this point, Eldor and coworkers demonstrated a reduced PGI_2 production after exposing irradiated cells to PGH_2, the substrate converted to PGI_2 by prostacyclin synthetase. They also demonstrated that augmentation of phospolipase A_2 activity by melittin and ionophore A23187 was reduced over time after irradiation (3).

Friedman and coworkers studied bovine pulmonary artery endothelial cells and related the radiation-induced increase of PGI_2 production to increased sodium-dependent transport of alphaaminoisobutyrate, a non-metabolizable amino acid (7). They believe that radiation depletes the intracellular arachidonate stores which inhibit sodium-dependent amino acid transport. Friedman et al. also contend that the nature of endothelial cell binding to its extracellular matrix and the subsequent alteration of the cytoskeleton modifies PGI_2 production and amino acid transport. They observe that cells more strongly adhered to a growth surface do not increase prostaglandin production and have reduced amino acid uptake after irradiation but cells more loosely adhered have increased production and uptake. However, they indicate that these findings require further clarification.

PROSTAGLANDIN PRODUCTION, ACUTE CELL VIABILITY AND PROLIFERATIVE SURVIVAL

My coworkers and I tried to determine whether enhancement of PGI_2 production by irradiation was associated with endothelial cell detachment from a growth surface (4). The hypothesis was that following irradiation, attached (viable) cells might have increased capacity to make the prostaglandin and detached cells might not. In our cultures, attached cells exclude Trypan Blue and detached cells do not. However, as indicated by Figure 3, detached irradiated cells also had increased capacity for PGI_2 production. This observation implies that capacity for increased prostaglandin production is not closely linked to cell viability following irradiation.

There is indirect evidence that alterations in incremental PGI_2 production are associated with reduced proliferative ability of endothelial cells. Hahn and coworkers observed a radiation-induced mitotic arrest simultaneous with an increased capacity of PGI_2 production in cultures not disrupted by enzymatic removal of cells (5). They interpreted these events as evidence that the boost in prostaglandin production was a signal to halt mitosis. Eldor and coworkers (3), on the other hand, related the reduced PGI_2 production to reduced colony formation observed after removal of cells from irradiated cultures.

PROSTAGLANDIN PRODUCTION, RECEPTORS AND CELL/CELL INTERACTIONS

Three areas of cell/cell interaction and its relationship to prostaglandins have been studied after irradiation, 1) the influence of one cell type to modify cell proliferation of another, 2) cell adhesion, and 3) modification of cell receptors.

PGE production by marrow stromal cells has been shown to stimulate erythroid colony formation (2). Gibson and colleagues demonstrated that the reduction of PGE synthesis by low dose radiation (1-8 Gy) was also related to reduced erythroid colony formation for cells innoculated onto a feeder layer of stromal cells (2). Conversely, as the marrow stromal

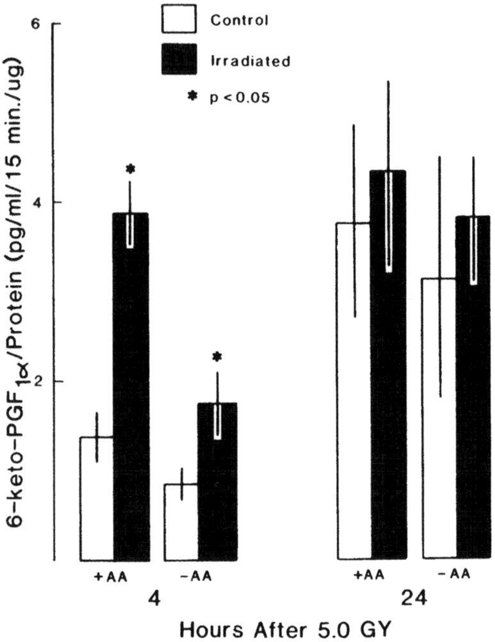

FIG. 3. Prostacyclin (PGI_2) production during a 15-min incubation of detached cells as a function of time after sham-irradiation (open bars) or 5.0 Gy of ^{60}Co γ-rays (solid bars). Cells were incubated in presence (+AA) or absence (-AA) or arachidonate (10 μM). PGI_2 was measured by radioimmunoassay of its stable metabolite 6-ketoprostaglandin $F_{1\alpha}$ (6-keto-$PGF_{1\alpha}$), and was corrected for protein content of the detached cells. Data are means ± SE, n = 6. (Reprinted with permission, J. Appl. Physiol. 58, 592-597, 1985).

cells produced more PGE when they were lysed by higher doses of radiation (100-200 Gy), there was increased erythroid colony formation.

Increased platelet (4) and leukocyte (9) adhesion to irradiated endothelial cell monolayers has been recently documented and PGI_2 may inhibit this process. However, only platelet adhesion has been studied stimultaneous to measurement of prostaglandin production following irradiation (4). In this study, no inhibition of platelet adhesion was observed despite elevation of PGI_2 production. A limitation of this observation is that the platelets were not activated by the addition of agents such as thrombin.

Irradiated cells may have a reduction of receptors for prostaglandins as indirectly demonstrated by an examination of IgM rheumatoid factor made by cultured human T lymphocytes (10). Ceuppens and coworkers observed that non-steroidal anti-inflammatory agents inhibit production of IgM rheumatoid factor and that by adding PGE_2 the inhibition was prevented. When lymphocytes were irradiated (10 Gy) prior to culture and examined 8 days later, the anti-inflammatory agent, indomethacin, did not inhibit IgM rheumatoid factor production. The authors speculated that PGE_2 receptors are expressed on suppressor T-cells and that endogenously produced PGE_2 exerts a tonic inhibition of a radiosensitive suppressor T-cell or that PGE_2 stimulates radiosensitive helper T cells. The major limitations of this speculation is that PGE_2 receptors were not directly examined nor were indicators of radiation-induced cytotoxicity studied. The phenomenon also could be explained, as the authors mention, by destruction of a subpopulation of T-cells that were more radiosensitive than others.

Vegt and colleagues also indirectly studied the effect of radiation on prostaglandin receptors in cultured endothelial cells from human umbilical vein (11). They found that 48 hours after 20 Gy PGE_1-stimulated c-AMP synthesis was reduced but could not conclude whether this reduction was due to decreased receptor binding, decreased interaction between receptor and adenylate cyclase or decreased adenylate cyclase activity.

EICOSANOIDS AND RADIOPROTECTION OF CULTURED CELLS

PGE compounds have been shown to be radioprotective in vivo (12) and a few investigators have tried to clarify this observation on cultured cells. Prasad first reported (13) that PGE_1 (10 μg/ml) added 1 or 4 hours prior to irradiation increased the survival of irradiated Chinese-hamster ovary cells by about a factor of 2. His end point was the number of colonies formed 12 days after 5 Gy. The addition of PGE_1 after irradiation did not enhance survival. He speculated that the protective mechanism of PGE_1 was from elevation of cyclic AMP, since a phospodiesterase inhibitor also enhanced colony formation in irradiated cells to a similar degree. However, cyclic AMP was not assayed.

A more thorough examination of this subject was reported by Lehnert (14) who studied colony formation in Chinese hamster V-79 cells. The main goal of this study was to elevate the intracellular level of cyclic AMP which was measured by radioimmunoassay. She verified that PGE_1 elevated cyclic AMP and that treatment with the prostaglandin 1 hour prior to irradiation also enhanced cell survival as measured by colony forming assays. However, Lehnert found that enhancement of colony formation was dependent on the dose of radiation. At low doses the shoulder of the survival curve was increased and more drug-treated cells survived low doses than did controls. However, the Do derived from the exponential slope of the survival curve was greater for control than for drug-treated cells, so that with increasing doses the latter became progressively more sensitive to irradiation. The survival curves for control and prostaglandin-treated cells crossed over at about 12 Gy. These results indicated that PGE_1 treatment increased the capacity of these cells to accumulate sublethal damage. Further experiments using two split doses of 6 Gy (1 to 3 hours apart) verified that cells treated with PGE_1 repaired sublethal damage 3- to 4-fold greater than untreated cells. Similar to Prasad's observation, phosphodiesterase inhibitors had an equivalent radioprotective effect to that of PGE_1, which was associated with elevation of intracellular cyclic AMP.

In contrast to the above observations, Millar and Jinks (15) did not find the prostaglandins PGE_1 and PGA_1 to be radioprotective for Chinese hamster cells (V79-753B). Their experiments differed from Prasad's and Lehnert's in two respects: a lower dose of prostaglandin was used for most experiments (1 μg/ml); and, perhaps more importantly, the length of prostaglandin exposure prior to irradiation was longer ("overnight" to 3 days).

Millar and Jinks also examined inhibition of prostaglandin synthesis in two cell lines (15,16), the Chinese hamster cells and cultured mouse fibrosarcoma cell $HSDM_1C_1$ which secrete 2 μg PGE_2/mg cell protein 24/hr. They found that inhibition of prostaglandin production with non-steroidal anti-inflammatory agents flurbiprofen and indomethacin did not affect the dose survival curve of these cells.

Walden and associates have recently reported on the radioprotective effect of leukotrienes in Chinese hamster V79A03 cells (15). Preliminary cell survival studies have shown a small but statistically significant decrease in cellular radiosensitivity when the cells are treated with LTD_4 2 hours before exposure to 10 Gy X-irradiation.

FUTURE DIRECTIONS

The cell biology of eicosanoids and the effects of gamma irradiation is a science in its nascency. Few cyclooxygenase products have been examined yet in irradiated cell cultures. Furthermore, there has been a paucity of cell types examined.

As markers of radiation-induced cytotoxicity in culture, a number of issues need clarification. Although eicosanoid production can be used to monitor acute cell damage, the significance of increased eicosanoid production initially and decreased production later is unknown. One can presume that alterations in eicosanoid metabolism after irradiation reflect pertubations to the cell membranes, but we do not know what other membrane changes are associated with this alteration, and only 2 reports on sodium-dependent amino acid transport have begun to address this issue (7,11). For example, is cell swelling, increased membrane rigidity, or abnormal calcium transport manifestations of the same process that drives and then retards prostaglandins production? Is the reduced prostaglandin production observed at later times following irradiation a protective or maladaptive response to the injury? We also suspect but do not know whether changed production of prostaglandins play a role in cell/cell interactions. For example, is the initial increase in prostaglandin production a stimulatory, inhibitory or neutral signal for chemotaxis, mitosis, and thrombosis?

A leading controversy in the radioprotective role of prostaglandins for cultured cells appears to be the length of both drug exposure and whether endogenous production of prostanoids are protective. The existing studies imply that short boosts in prostaglandin concentrations is more radioprotective than prolonged exposure but this issue still has to be tested rigorously.

Answers to the above questions should provide direction for the more complicated experimental approaches in vivo. Advances in testing single cells, the development of co-culture techniques of 2 or more cell types, and improved simulation of in vivo conditions are promising approaches for many of the issues raised by previous work on cultured cells.

BIBLIOGRAPHY

1. P. Polgar, G. Hahn, and L. Taylor, Radiation Effects on Eicosanoid Formation. In, Biochemistry of Arachidonic Acid Metabolism (WEM Lands, Ed.) pp. 161-173. M. Nijhoff, Boston, 1985.

2. D. P. Gibson, R.L. Degowin, and S.A. Knapp, Effect of X Irradiation on Release of Prostaglandin E from Marrow Stromal Cells in Culture. Radiat. Res. 89, 537-545 (1982).

3. A. Eldor, I. Vlodavsky, E. HyAm, R. Atzmon, and Z. Fuks, The Effect of Radiation on Prostacyclin (PGI_2) Production by Cultured Endothelial Cells. Prostaglandins 25, 263-279 (1983).

4. D.B. Rubin, E.A. Drab, C. Ts'ao, D. Gardner, and W.F. Ward, Prostacyclin Synthesis in Irradiated Endothelial Cells Cultured from Bovine Aorta. J. Appl. Physiol. 58, 592-597 (1985).

5. G.L. Hahn, M. Menconi, and P. Polgar, The Effect of Gamma Radiation on Prostacyclin Production in Cultured Pulmonary Artery Endothelium. In, Prostaglandins and Cancer: First International Conference (T.J. Powles, R.S. Bockman, K.V. Honn, and P. Ramwell, Eds.), pp. 381-384. Alan R. Liss, Inc., New York, 1981.

6. G.L. Hahn, M.J. Menconi, M. Cahill, and P. Polgar, The Influence of Gamma Radiation on Arachidonic Acid Release and Prostacyclin Synthesis. Prostaglandins 25, 783-791 (1983).

7. M. Friedman, D.S. Saunders, M.C. Modden, E.L. Chaney, and L. Kwock, The Effects of Ionizing Radiation on the Pulmonary Endothelial Cell Uptake of α-Aminoisobutyric Acid and Synthesis of Prostacyclin. Radiat. Res. 106, 171-181 (1986).

8. M.E. Hemler, H.W. Cook, and W.E.M. Lands, Prostaglandin Biosynthesis Can Be Triggered by Lipid Peroxides. Arch. of Biochem. and Biophys. 193, 340-345 (1979).

9. M.M. Dunn, E.A. Drab, and D.B. Rubin, Effects of Irradiation on Endothelial Cell-Polymorphonuclear Leukocyte Interaction. J. Appl. Physiol. 60, 1932-1937 (1986).

10. J.L. Ceuppens, M.A. Rodriguez, and J.S. Goodwin, Non-steroidal Anti-inflammatory Agents Inhibit the Synthesis of IgM Rheumatoid Factor In Vitro. Lancet 1 (9271), 528-530 (1982).

11. G.B. Vegt, A.M. Wassenaar, E.W.M. Kawilarang-DeHaas, P.O. Schutte, M. Van der Linden, M. Di Bon-de Ruijter and A. Boon, Radiation-induced Changes in Cell Membrane of Cultured Human Endothelial Cells. Radiat. Res. 104, 317-328 (1985).

12. W.R. Hanson and C. Thomas, 16, 16-Dimethyl Prostaglandin E2 Increases Survival of Murine Intestinal Stem Cells when Given Before Photon Radiation. Radiat. Res. 96, 393-398 (1983).

13. K.N. Prasad, Radioprotective Effect of Prostaglandin and an Inhibitor of Cyclic Nucleotide Phosphodiesterase on Mammalian Cells in Culture. Int. J. Radiat. Biol. 22, 187-189 (1972).

14. S. Lehnert, Modification of Postirradiation Survival of Mammalian Cells by Intracellular Cyclic AMP. Radiat. Res. 62, 107-116 (1975).

15. B.C. Millar and S. Jinks, Do Prostaglandins Affect Cellular Radiosensitivity? Int. J. Radiat. Biol. 46, 367-373 (1984).

16. B.C. Millar, S. Jinks and T.J. Powles, Flurbiprofen, a Nonsteroidal Anti-inflammatory Agent, Protects Cells Against Hypoxic Cell Radiosensitizers In Vitro. Br. J. Cancer 44, 733-740 (1981).

17. T. L. Walden, Jr., E.V. Holahan, Jr., and G.N. Catravas, Development of a Model System to Study Leukotriene-induced Modification of Radiation Sensitivity in Mammalian Cells. Prog. Lipid Res. 25, 587-590 (1986).

2

EFFECTS OF GAMMA IRRADIATION ON PROSTAGLANDIN BIOSYNTHESIS AND METABOLISM: INTACT TISSUES.

DANIEL M. STRICKLAND, MD and MURRAY D. MITCHELL, D PHIL

Division of Maternal & Child Health
Wilford Hall USAF Medical Center, San Antonio, Texas 78236
and
Department of Reproductive Medicine
University of California-San Diego, La Jolla, California 92093

INTRODUCTION

Exposure to gamma irradiation can affect the synthesis and metabolism of potent biochemical agents within various organs, tissues, and cell types, resulting in a wide range of effects (1,2), potentially either harmful or protective. Ionizing radiation has been shown to produce changes in prostaglandin production by a variety of tissues (3,4); however the precise mechanisms of such alterations and the role of prostaglandins in the patho-physiological effects of ionizing radiation have not been fully elucidated.

Various methods, discussed in detail elsewhere in this volume, have been employed to study the effects of radiation on prostaglandin biosynthesis and metabolism. The advantages and limitations of these methods must be considered when evaluating the results of any investigation.

Subcellular fractionation

At the subcellular level specific biochemical processes can be evaluated, examples of which include the effects of radiation on plasma membrane prostaglandin receptors, cytosolic and microsomal enzyme degradation from free radical production, or changes in DNA/RNA that alter prostaglandin biosynthesis or metabolism. Mechanisms that require dynamic interaction between subcellular organelles cannot be readily examined at this level.

Cell culture

Cells in suspension or in monolayer culture are a popular and convenient preparation for the study of cellular processes. Because of transformations occurring during growth, however, cells maintained in culture may bear little structural or functional resemblance to their parents. Moreover, the important relationships between epithelial elements and mesenchyme are disrupted by this method. Platelets are a special type of cell with generally intact cytosolic processes but without nuclei or the capacity to replenish prostaglandin synthesizing enzymes.

Intact tissue

The next level of organization is intact tissue, that may be irradiated *in vivo* and then removed for analysis, or may be studied with *in vitro* organ culture. Prostaglandin concentrations measured in irradiated sections of isolated tissues or in tissues from animals exposed to whole body irradiation are the result of synthesis by several cell types. The cells maintain relationships with contiguous cells and their potentially mediating influences; this is not so with cells in culture.

Whole animals

Irradiating whole intact animals, although feasible even for large animals, is cumbersome. More importantly, cumulative whole body effects may be difficult to distinguish from effects on specific organs or tissues. Because the blood level of a prostaglandin reflects whole body effects, a change in prostaglandin production by one organ may not be distinguishable or may be obscured by unrelated or reciprocal changes elsewhere.

Furthermore, circulating concentrations of most primary prostaglandins are, in general, too low to be determined; thus metabolites must be quantified. Changes in plasma concentrations of metabolites are only indirectly related to changes in prostaglandin production and may be reflective of combinations of changes in synthesis, metabolism, and clearance.

15

SPECIFIC TISSUES

Blood vessel

This tissue type is covered in detail in a separate chapter of this volume. In general, however, the principal prostaglandin produced by vascular endothelium is prostacyclin (PGI_2) (5,6), which has a critical role in several aspects of circulatory physiology, including inhibition of platelet aggregation and relaxation of vascular smooth muscle. Thus any agent or insult that affects endothelial prostacyclin production can produce significant alterations in vascular homeostasis. Pathological changes in blood vessels after exposure to radiation include morphological alteration of the intima, intraluminal and perivascular edema, and the formation of vaso-occlusive thrombi (2,7). Radiotherapy also has been found to act synergistically with other risk factors to increase the frequency of atherothrombotic complications after irradiation (7,8).

Hahn, Menconi, and Polgar (5) demonstrated that upon exposure to gamma irradiation, endothelial cells cultured from pulmonary artery cease growth but attain an increased capacity to synthesize prostacyclin. Further investigation by this group (6) revealed that the mechanism for this increased synthetic capacity appears to be both by enhancement of release of arachidonic acid (AA) from lipid stores and by increased conversion of AA to prostacyclin. Eldor and coworkers (9), in a detailed study reporting somewhat different data, concluded that endothelial cells in culture exposed to high doses of radiation (1000-5000 rad) released prostacyclin because of cellular damage. Lower doses resulted in a marked decrease in the capacity to synthesize prostacyclin.

Reports on the effect of gamma irradiation on prostaglandin synthesis by intact vascular tissue have resulted in similarly mixed conclusions. Allen et al (10) demonstrated a significant decrease in prostacyclin production by intact human umbilical artery after exposure to a single 200 rad dose of gamma irradiation. Platelet thromboxane production was not inhibited, however, by doses as high as 2000 rad. Sinzinger (11) reported a biphasic effect of gamma irradiation on prostacyclin production by rabbit aorta. The initial effect, peaking at 2-4 hours after irradiation, was a significant, but temporary, increase in prostacyclin production that was postulated to result from damage to endothelial cells. Forty eight hours after exposure, however, prostacyclin production had declined significantly compared with control production.

Lung

Eisen and Walker (3,4) provided some of the earliest data on the effects of ionizing radiation on prostaglandin-like activity in specific tissues. In their experiments, mice were exposed to whole body x-ray doses of 200 to 700 rad. The mice were sacrificed at variable times after exposure and tissue prostaglandin-like activity was assessed; specific prostaglandins were not quantified, however.

In the lung, prostaglandin-like activity fell slightly for the first two days after exposure, then increased to a transient peak 4 days after treatment. After one week, however, the activity was slightly lower than in non-irradiated lungs. The effects were noted only at exposures of 500 rad or more and no changes in lung weights were noted.

Steel and Catravas (12), in an extensive study, irradiated guinea pigs with gamma radiation in doses from 500 to 3000 rad and sacrificed the animals at variable times after exposure. Special care was taken to flush all blood from the lungs before they were removed from the animal for processing. The tissue was then perfused and incubated in Tyrode's buffer and basal and stimulated production of prostaglandins E, F, and thromboxane B_2 determined.

Rates of production of prostaglandins E and F (PGE and PGF) increased significantly at 1 to 3 hours after exposure but by 6 hours production had returned to control levels. Thromboxane B_2 production was significantly higher than controls in all irradiated animals and production rates remained elevated for up to 48 hours after irradiation. The effects on all prostanoids were generally dose related.

Heinz and co-workers (13) investigated the effects of whole body gamma irradiation on thromboxane B_2 (TXB_2) and 6-keto-prostaglandin $F_{1\alpha}$ (6-keto-$PGF_{1\alpha}$) synthesis in isolated, perfused rat lungs. At variable times after irradiation, the lungs were cleared of blood and placed into a perfusion chamber. Lung expansion was maintained with a respirator and the pulmonary artery was perfused with Krebs-Ringer bicarbonate buffer and prostanoid concentrations determined in the perfusate.

Lungs exposed to doses as low as 500 rad demonstrated a significant increase in TXB_2 concentrations in the perfusate 12 hours after exposure; by 120 hours after exposure levels had decreased significantly compared to non-irradiated controls. Lung perfusate 6-keto-$PGF_{1\alpha}$ concentrations had increased significantly within 12 hours of exposure to 500 rad but by 120 hours had also decreased significantly compared with controls.

Brain

Central nervous irradiation, for conditions such as leukemia/lymphoma or brain tumors has been associated with alterations in vascular permeability (14) and the development of a mineralising microangiopathy with dystrophic calcifications in the brain (15,16). Eisen and Walker, however, detected no change in prostaglandin-like activity in brains of mice exposed to up to 700 rad of x-irradiation. This may be reflective of the relative radio-resistance of neuronal tissue (2), although the effects of radiation on the CNS may be mediated primarily through actions on blood vessels. Thus, as a function of total tissue weight, radiation-induced changes in prostaglandin production may be too small to detect. It is possible also that changes in the production of one prostaglandin may be countered by reciprocal changes in the production of another prostaglandin, the result of which could not be detected by measuring only total prostaglandin-like activity.

Consistent with this hypothesis, Pausescu and coworkers (17) reported that gamma-irradiation of rats with 500 and 750 rad induced an increase in PGE content of brain tissue and a decrease in PGF content. Thus, an increase in the synthesis of a vasodilatory prostaglandin (PGE) was offset in a quantitative sense by a decrease in the production of a vasoconstrictive prostaglandin (PGF). The homeostatic response to such changes might be vasodilatation, a potentially protective response, resulting in increased blood flow, improved tissue oxygenation, and enhanced removal of toxic substances produced by radiation damage.

Kidney

A transient decrease in prostaglandin-like activity in the kidney of x-irradiated mice at 2 days after exposure has been reported (3,4); by day 4, however, the tissue activity was normal. Interestingly, although significant changes with time could be demonstrated for exposed animals, the actual prostaglandin-like activity was not significantly different from the activity detected in non-irradiated animals.

Liver and Spleen

Eisen and Walker (3,4) determined prostaglandin-like activity in the spleens of mice exposed to x-rays at doses of 200 to 700 rad. Prostaglandin-like activity increased significantly and acutely at all radiation doses used. The prostaglandin-like activity corresponded to that of both authentic PGE and PGF on thin layer chromatography. The same authors reported also that irradiation promptly and potently reduced the activity of 15-hydroxy-prostaglandin dehydrogenase (PGDH), a key prostaglandin metabolizing enzyme.

Trocha and Catravas (18,19) reported a transient rise in PGF content in rat liver and spleen at 3-6 hours after the administration of 1000 rad gamma-irradiation; at 4 to 7 days PGF levels rose markedly. PGE levels did not show an early rise but at 4 and 7 days post-irradiation levels in liver and spleen of irradiated animals were significantly higher than in controls. The authors also demonstrated significant increases in liver beta-glucuronidase activity after irradiation, the changes paralleling changes in prostaglandins. They interpreted such concomitant changes as indicative of either lysosomal damage resulting in increased prostaglandin production or of prostaglandin mediated activation of lysosomes.

Pausescu, et al (17) found no effect on hepatic PGF production in rats exposed to 500 rad gamma-radiation but a transient increase in animals exposed to 750 rad. PGE production was increased significantly up to 18 days after irradiation at both doses.

Gut

There is circumstantial evidence that is supportive of the involvement of prostaglandins in radiation-induced enteritis (20,21). Eisen and Walker (3,4) noted considerable variation between test animals in their examination of prostaglandin-like activity in the small intestine of irradiated mice. An initial fall in activity after exposure to x-irradiation and a subsequent rise by day seven were not statistically significant. Borowska (22) on the other hand, in a study also of irradiated mice, found that prostaglandin-like activity increased significantly in the small intestine and remained elevated until 4 days after exposure. The increase in prostaglandins, as well as a corresponding increase in small intestine motility, were inhibited by pre-treatment with indomethacin.

It is important to note that in both studies cited above the intestines were removed after irradiation and the tissues were homogenized for extraction of prostaglandin-like activity. Even though both groups of investigators performed the homogenation process in ice-cold saline to which indomethacin had been added, the amount of prostaglandins produced by the homogenation process itself may have been appreciable. Thus, it is not possible from these studies to conclude whether the prostaglandin-like activity represented an increased capacity to synthesize prostaglandins *in vivo* or enhancement of prostaglandin biosynthesis in response to tissue trauma.

Gal et al (23), using a technique of tissue superfusion, evaluated the effects of in vitro irradiation on the production of prostaglandins by human colon. The technique utilized a "washout" period in an attempt to eliminate prostaglandins formed during the tissue preparation, subsequent prostaglandin synthesis being more reflective of in vivo synthetic capacity, than of trauma artifact.

Squares of tissue in organ culture dishes were gamma irradiated with 500, 1000, or 2500 rad in single application; tissues that were not irradiated served as controls. The tissues in some of the dishes were treated during the preparation and radiation periods with mefenamic acid, an inhibitor of prostaglandin biosynthesis. After treatment the tissues were superfused (without inhibitors) and prostaglandin concentrations in the effluent fluid were determined by specific radioimmunoassay.

As shown in Fig 1, both mefenamic acid and 2500 rad inhibited significantly the production of PGE_2, $PGF_{2\alpha}$, PGD_2, 6-keto-$PGF_{1\alpha}$, thromboxane B_2, and PGFM (13,14-dihydo-15-keto-$PGF_{2\alpha}$), the latter three prostanoids being stable metabolites of prostacyclin, thromboxane A_2, and prostaglandin $F_{2\alpha}$, respectively. Fig 2 illustrates a dose-response effect of gamma irradiation on PGE_2 and $PGF_{2\alpha}$ production.

Lactate dehydrogenase (LDH) activity in the effluent was also assayed to determine whether cell viability decreased during the period of superfusion. After a decrease in LDH activity released during the washout period, there was no significant change in LDH activity released from any group of specimens over a period of 1.5 hours of superfusion. Furthermore there was no significant difference between LDH activity released from irradiated and nonirradiated bowel. The ratio of PGFM:$PGF_{2\alpha}$ did not differ significantly between irradiated and non-irradiated bowel, suggestive that there was no increase in PGDH metabolizing activity as a result of irradiation.

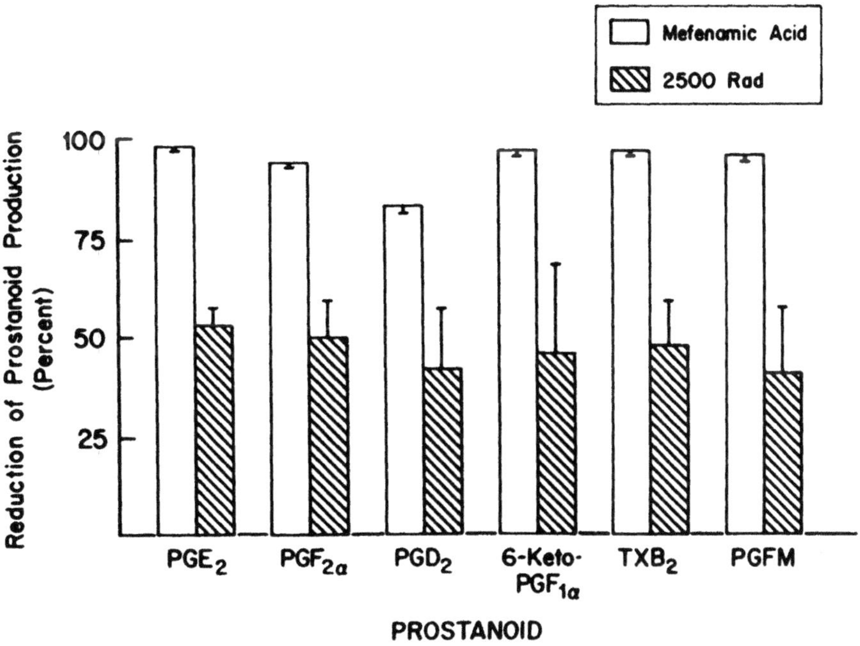

Figure 1. Comparison of the effect of mefenamic acid (100 mcg/ml) and irradiation (2500 rad) on prostanoid production by human bowel in superfusion. The reduction in prostanoid production is expressed as the mean (± S.D.) percent reduction from rates of prostanoid production by control (nonirradiated) bowel, assayed in triplicate collections from one experiment. (Reprinted with permission from ref. 23)

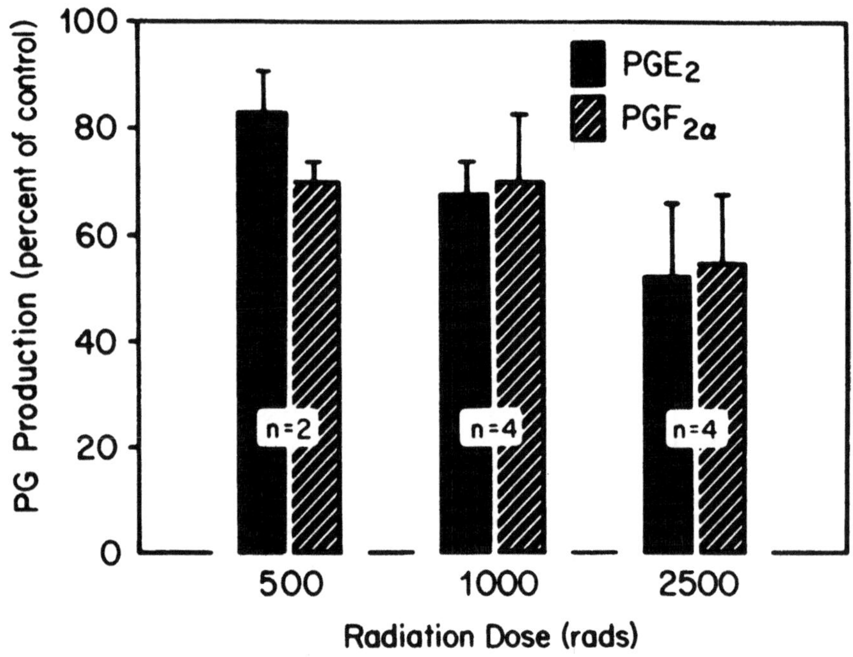

Figure 2. The effect of radiation on PGE_2 and $PGF_{2\alpha}$ production by human bowel. The data are expressed as a percentage (mean ± SEM) of the rates of production of control (nonirradiated) bowel during the first 90 minutes of washout. Prostaglandins were assayed in triplicate collections from 4 experiments in which bowel was irradiated with 1000 and 2500 rads, and in 2 experiments in which bowel was irradiated with 500 rads. Rates of production by tissues receiving 1000 and 2500 rads were significantly less ($p < 0.05$) than nonirradiated (control) tissues. (Reprinted with permission from ref. 23)

The authors concluded that radiation caused an acute decrease in prostaglandin biosynthesis in a dose-dependent manner and that this effect was not the result of either increased early cell death or decreased prostaglandin catabolism.

SUMMARY AND CONCLUSIONS

The reported effects on prostaglandin biosynthesis and metabolism of exposing intact tissue to ionizing radiation have been variable. The few studies that have addressed this issue have utilized methodologies that are difficult to compare, have quantified different prostaglandins, and have made such determinations at different times after irradiation. Thus a unified picture of prostanoid biosynthetic responses in tissue to irradiation remains elusive.

From the available data, however, it is almost certain that such responses are tissue-dependent and prostanoid specific. This conclusion is reasonable in an *a priori* sense also, because the functions of various tissues are specific and the biologic effects of different prostanoids also are highly specific.

An appreciation for the potential roles for manipulation of prostaglandin synthesis in managing radiation therapy requires much more data, both basic and clinical, than are currently available. Investigators, however, must be mindful of the significant methodological problems encountered in the investigation of such volatile and ephemeral substances as the prostaglandins.

The opinions expressed in this manuscript are those of the authors and not necessarily those of the United States Air Force or the Department of Defense.

REFERENCES

1. Altman KI, Gerber GB, Okada S: Radiation Biochemistry, Vol II:Tissues and body fluids. Academic Press, New York, 1970
2. Berdjis CC: Pathology of Irradiation. Williams and Wilkins, Baltimore, 1971
3. Eisen V and Walker DI: Br J Pharmac 57:527-532, 1976
4. Eisen V, Walker DI, Binysh SG, Tedder RS: Agents and Actions, Suppl 2:99-103, 1977
5. Hahn GL, Menconi MJ, Polgar P: In: Prostaglandins and Cancer (eds. TJ Powles, RS Bockman, KV Honn, P Ramwell) Alan R. Liss, New York, 1981, pp. 381-384
6. Hahn GL, Menconi MJ, Cahill M, and Polgar P: Prostaglandins 25:783-791, 1983
7. Rubin P, Casarett GW: In: Clinical Radiation Pathology, Vol I (eds. P Rubin , GW Casarett) W. B. Saunders, Philadelphia 1968, pp. 38-61
8. Dunsmore LD, LoPonte MA, Dunsmore RA: J Am Coll Cardiol 8:239-244, 1986
9. Eldor A, Vlodavsky I, HyAm E, Atzmon R, Fuks Z: Prostaglandins 25:263-279, 1983
10. Allen JB, Sagerman RH, Stuart MJ: Lancet 2:1193-1196, 1981
11. Sinzinger H, Firbas W, Cromwell M: Prostaglandins 24:323-329, 1982
12. Steel LK and Catravas GN: Int J Radiat Biol 42:517-530, 1982
13. Heinz TR, Schneidkraut MJ, Kot PA, Ramwell PW, Rose JC: Prog Biochem Pharmacol 20:74-83, 1985
14. Haymaker W, Ibrahim MZM, Miguel J, Cell N, Riopelle AJ: J Neuropathol Exp Neurol 17:50-79, 1958
15. Rubenstein LJ, Herman MM, Long TF, Wilbur JR: Cancer 35:291-305, 1975
16. Price RA, Birdwell DA: Cancer 42:717-728, 1978
17. Pausescu E, Chirvasie R, Teodosiu T, Paun C: Rad Res 65:163-171, 1976
18. Trocha PJ and Catravas GN: Int J Radiat Biol 38:503-511, 1980
19. Trocha PJ and Catravas GN: In: Advances in Prostaglandin and Thromboxane Research, Vol. 7 (eds. B Samuelsson , PW Ramwell, and R Paoletti), Raven Press, New York, 1980, pp. 851-855
20. Tanner NSB, Stamford IF, Bennet A: Br J Cancer 43: 767-771, 1981
21. Northway MG, Libshitz HI, Osborne BM, Feldman MS, Mamel JJ, West JH, Szwarc ILA: Gastroenterology 78:883-892, 1980
22. Borowska A, Sierakowski S, Mackowiak J, Wisniewski K: Experientia 35:1368-1370, 1979
23. Gal D, Strickland DM, Lifshitz S, Buchsbaum HJ, Mitchell MD: Int J Rad Biol Oncol Phys 10:653-657, 1984

3

RADIATION-INDUCED ALTERATIONS IN CYCLOOXYGENASE PRODUCT RELEASE: AN OVERVIEW

M.J. Schneidkraut, P.W. Ramwell and P.A. Kot

Department of Physiology and Biophysics, Georgetown University Medical Center, 3900 Reservoir Road, N.W., Washington, D.C. 20007

Cyclooxygenase products, by virtue of their vasoconstrictor (PGF2a and TXA2) (1-5) and vasodilator actions (PGE2 and PGI2) (1-3,6-9) as well as their platelet aggregatory (TXA2, PGG2 and PGH2) (10-14) and anti-aggregatory properties (PGE2 and PGI2) (15-18) may play a role in radiation-induced tissue injury. This hypothesis is based on the observation that radiation exposure results in tissue injury (19-35) and tissue injury is associated with the release of cyclooxygenase products (36-39).

In order to determine if cyclooxygenase products play a role in radiation-induced tissue injury, an association between radiation exposure and altered cyclooxygenase product release must first be demonstrated. Secondly, an evaluation of the effect of cyclooxygenase pathway inhibitors and exogenous prostaglandins on radiation-induced tissue injury must be made. This review is divided into two sections: 1) radiation-induced altertions in cyclooxygenase product release and 2) the role of cyclooxygenase products in radiation-induced tissue injury. An overview of the literature will be presented both in support of, and against, the hypothesis stated above.

I. RADIATION-INDUCED ALTERATIONS IN CYCLOOXYGENASE PRODUCT RELEASE

A. Cellular and parenchymal tissue release: Early studies evaluated the release of the classical prostaglandins, PGE2 and PGF2a, by various organs following whole body ionizing radiation exposure (5,9,32,40-50). In one study, Eisen and Walker (41) exposed mice to a dose of 2 to 7 Gy whole body irradiation. At varying time intervals (0 to 7 days) post irradiation, the animals were re-anesthetized, exsanguinated, and the lungs and spleen removed. The organs were homogenized, the

P. Polgar (ed.), EICOSANOIDS AND RADIATION.

prostaglandins extracted, and the PGE content quantitated by the rat stomach strip bioassay (41). Lung homogenates showed an increase in PGE-like activity four days after exposure to six or seven Gy whole body irradiation. Splenic homogenates also showed increased PGE-like activity, though the increase occurred earlier (2 to 7 days post irradiation) and with a smaller (2 Gy) dose of radiation (41). A subsequent study by Walker and Eisen (46) re-confirmed the effect of whole body irradiation on pulmonary and splenic PGE-like activity but also showed that altered cyclooxygenase product release occurred earlier than previously demonstrated. In this second study, a transient increase in splenic PGE occurred within four hours of exposure (46). The jejunum also released increased amounts of PGE four days after radiation exposure (51).

Trocha and Catravas (45,48) extended these observations by determining immunoassayable PGE and PGF2a release from spleen and liver homogenates. Hepatic iPGE and iPGF2a were increased three to six hours post irradiation. Both prostaglandin concentrations decreased to control within nine hours, but hepatic iPGF2a levels were again elevated at four and seven days post irradiation. These investigations confirmed the earlier studies of Eisen and Walker (41,46) by showing an increase in splenic iPGE 3 hours, 3 to 4 days, and 7 days post irradiation. The splenic iPGF2a levels paralleled the increased iPGE release.

The earliest observed change in classical prostaglandin release following whole body irradiation was demonstrated by Nikandrova et al (42). Mice were exposed to 9 Gy whole body irradiation and the liver, brain and testis removed at various times after radiation exposure. Significant increases in both PGE2 and PGF2a levels were observed as early as 15 minutes post irradiation. Unlike the investigations of Trocha and Catravas (45,48), Nikandrova et al (42) could not demonstrate an increase in hepatic PGE2 or PGF2a between one and 72 hours post irradiation. However, elevations in classical prostaglandin values were observed in both the testis and brain homogenates. These studies indicate that radiation exposure, at doses capable of inducing tissue injury, alters the release of PGE2 and PGF2a from various organs. The time interval between irradiation and increased cyclooxygenase product release, as well as the minimum doses of radiation necessary to alter

PGE2 and PGF2a release was variable depending on the organ evaluated and the species studied.

The effect of radiation exposure on TXA2 release has also been evaluated (39,43,44,49,51-54). Allen et al (51) exposed human platelet rich plasma (PRP) to 0, 2, or 20 Gy ionizing radiation in vitro. The platelets were then challenged with ^{14}C-labelled arachidonic acid. The amount of ^{14}C-labelled TXB2 synthesized was measured. Irradiated and non-irradiated PRP released comparable amounts of TXB2 (51). Thus, in platelets, TXB2 synthesis from exogenous arachidonic acid is apparently unaffected by ionizing radiation exposure (51).

Maclouf et al (39) evaluated TXB2 release from rat splenic microsomes using gas chromatography-mass spectrometry. Rats were exposed to 9 Gy whole body irradiation and the spleens were removed 1 to 9 days post irradiation. TXB2 release from irradiated splenic microsomes was elevated 1 to 6 days post irradiation. Other studies determined the release of TXB2 from lung tissue (43,44). Guinea pigs were exposed to 0.5, 1.5, or 3.0 Gy whole body irradiation and lung parenchymal strips were assayed for basal TXB2 release over a 96 hour observation period by RIA (43). The basal release of iTXB2 was increased at 48 and 72 hours after 0.5 or 1.5 Gy doses of gamma irradiation. Exposure to a dose of 3.0 Gy, increased lung parenchymal iTXB2 release 3, 48, and 72 hours post irradiation (43). These studies suggest that the cyclooxygenase pathway in the lung and spleen is more radiosensitive than in platelets.

A subsequent study showed that the radiosensitivity of the cyclooxgenase pathway in different tissues within an organ varied. Guinea pigs were exposed to 3.0 Gy whole body irradiation or sham irradiation 1 to 336 hours before the lungs were removed (44). The airway tissue was dissected free of all surrounding parenchymal tissue and incubated at 37° for 45 minutes in a buffer solution. No radiation-induced alteration in iTXB2 synthesis was observed in the airway tissue.

Radiation-induced alterations in PGI2 release have been investigated and shown to be complex (51,52,54-61). Allen et al (51) irradiated umbilical arteries with 0.1, 0.25, or 1.0 Gy doses of radiation in vitro and challenged the vascular segments with ^{14}C-

labelled arachidonic acid. The incubation media was analyzed for the hydrolysis product of prostacyclin (6-keto prostaglandin F1a, 6KPGF1a) by RIA. Doses of radiation between 0.25 and 1.0 Gy significantly decreased i6KPGF1a release compared to sham irradiated vascular rings (51). These studies show that the vascular synthesis of 6KPGF1a from exogenous arachidonic acid, is suppressed by radiation exposure.

Basal release of PGI2 from irradiated tissue has also been assessed. Sinzinger et al (60) irradiated the abdominal aorta with 10 to 50 Gy gamma irradiation and measured vascular PGI2 synthesis by a platelet aggregation bioassay 1 to 6 months after irradiation. At each of the observation times, and at each dose of radiation, the abdominal aortic PGI2 synthesis was reduced compared to sham irradiated controls. On the other hand, bovine aortic endothelial cells irradiated with 0.5 to 5.0 Gy (59) or with 10.0 to 50.0 Gy (56) ionizing radiation had an increased i6KPGF1a release 24 hours post irradiation. Similar results were observed when pulmonary endothelial cells were exposed to either 4.0 to 20.0 Gy (58) or to 1.5 to 30.0 Gy (57) ionizing radiation.

The investigations described above indicate that irradiation of vascular tissue results in an increase or decrease in PGI2 release. These contradictory observations can be reconciled by the studies of Sinzinger et al (60,61). The increased release of i6KPGF1a occurred early after irradiation (within 24 hours) (61) and the decreased release of this same arachidonic acid metabolite occurred much later (longer than one month post irradiation) (60). Ts'ao et al (62) also showed a biphasic release of bioassayable PGI2 after irradiation, but these investigators demonstrated an early decrease in PGI2 release (24 hours post irradiation) followed by a prolonged increase in PGI2 values (six months after radiation exposure). These studies show that the release of PGI2 from irradiated tissue varies with the species studied and the time between irradiation and sample collection.

B. <u>Release from isolated perfused organs</u>: Studies from this laboratory have assessed the effect of whole body ionizing radiation exposure on cyclooxygenase product release from isolated perfused rat lungs and kidneys (52,54). Male rats were irradiated with 20 Gy gamma irradiation. Four hours later the lungs were removed and perfused *in vitro* with Krebs-Ringer bicarbonate plus 3% bovine serum albumin

(52). Pulmonary effluent from irradiated isolated perfused rat lungs showed an increased iTXB2 and i6KPGF1a release (99% and 50% respectively) compared to sham irradiated controls (52).

Radiation-induced alterations in cyclooxygenase product release by isolated perfused kidneys was also assessed (54). Rats were exposed to 20 Gy whole body irradiation four hours before the kidneys were isolated. Urine iPGE2 and i6KPGF1a excretion from irradiated isolated perfused kidneys was significantly greater than from sham irradiated kidneys (221% and 354% respectively) (54). The excretion rate of iTXB2 from irradiated isolated perfused kidneys was not significantly different than control (54).

Cyclooxygenase products were also determined in the renal venous effluent of irradiated and sham irradiated isolated perfused rat kidneys (54). Renal venous iTXB2 and i6KPGF1a outputs were comparable in both irradiated and sham irradiated isolated perfused kidneys (54). Renal venous iPGE2 values were not detectable in this system.

These studies show that following irradiation, isolated perfused organs exhibit an altered cyclooxygenase product release. These data also indicate that the pattern of release is variable from organ to organ. Finally, the radiation-induced alterations in cyclooxygenase product release is dependent on which cyclooxygenase product is assayed as well as the sample fluid assayed.

C. In vivo release: In a series of experiments from this laboratory, the effect of whole body radiation exposure on urine iTXB2 excretion was assessed (49,53). Male rats were exposed to either 0, 2, 10, or 20 Gy whole body irradiation and urine iTXB2 was measured 2 to 120 hours post irradiation (49,53). Urine iTXB2 was significantly elevated 4 to 120 hours after 10 Gy as well as 4 and 12 hours after 20 Gy radiation exposure (53). These results were confirmed by us a third time in a more recent series of investigations (54). In addition, the effect of whole body irradiation on iTXB2 excretion was corroborated by a second group of investigators (63,64).

The effect of whole body radiation exposure on urine iPGE2 excretion is less clear. Donlon et al (63,64) indicated a significant increase in iPGE2 excretion 6 to 96 hours after 9 Gy whole body irradiation (63). Recent studies from our laboratory suggest that exposure of rats to 20 Gy results in a significant decrease in iPGE2 excretion (54). This difference may be due, in part, to the different doses of radiation used. Further studies are necessary to clarify these data.

II. ROLE OF CYCLOOXYGENASE PRODUCTS IN RADIATION-INDUCED TISSUE INJURY

Several approaches have been used to evaluate the role cyclooxygenase products play in radiation-induced tissue injury. One such approach is to inhibit the radiation-induced increase in cyclooxygenase product synthesis using steroids (65-68) or non-steroidal anti-inflammatory drugs (26,66,69-73).

A. Steroids: Steroidal compounds inhibit membrane acyl hydrolase activity (37) which decreases the release of arachidonic acid from membrane phospholipids thereby decreasing cyclooxygenase product synthesis. Jacob et al (65) determined the effect of dexamethasone on the 30 day survival of mice exposed to 8.0 Gy whole body irradiation. Mice were divided into six treatment groups: 1) dexamethasone (0.6 mg/kg, i.p.) one hour before irradiation, 2) vehicle one hour before irradiation, 3) dexamethasone within 15 minutes after irradiation, 4) vehicle within 15 minutes after irradiation, 5) dexamethasone daily for four days after irradiation, and 6) vehicle daily for four days after irradiation. None of the dexamethasone treatment regimens altered the survival rate of the mice (65). On the other hand, Millar et al (67) pretreated a line of Chinese hampster cells (V-79-753B) with 1 ug/ml dexamethasone prior to exposure to 10 to 50 Gy ionizing radiation. Dexamethasone pretreatment increased the radioresistance of this cell line (67). A similar response was observed by Northway et al (73). Opossums were pretreated with 25 mg/kg hydrocortisone one hour before, and twice daily for one week following irradiation of the esophagus. Endoscopy and histology of the esophagus from steroid treated animals revealed significantly less tissue damage than saline treated controls (73).

It is difficult to draw conclusions about the role of cyclooxygenase products in radiation injury based on the results of the above mentioned studies because arachidonic acid metabolite levels were not determined. In addition, steroids have diverse actions besides their effect on membrane acyl hydrolases. A more specific approach to this problem would be to use non-steroidal anti-inflammatory drugs which inhibit cyclooxygenase activity (74-86).

B. Non-steroidal anti-inflammatory drugs: Millar and Jinks (66), as a follow up to their studies on the radioprotective effect of steroids, studied the effect of three cyclooxygenase inhibitors, indomethacin, benoxaprofen, or aspirin, on the radioresistance of Chinese hampster cells (V-79-753B). The cells were incubated with the cyclooxygenase inhibitor overnight prior to being exposed to graded doses of radiation (0 to 40 Gy) in vitro. Inhibition of cyclooxygenase activity did not increase the radioresistance of these cells (66). A previous study by this same group (68) evaluated the radioprotective effect of a fourth cyclooxygenase inhibitor, flurbiprofen, and came to the same conclusion. Thus, the radioprotective effect of steroids appears to be independent of their inhibitory actions on cyclooxygenase product synthesis.

Northway et al (73,26) also evaluated the effect of non-steroidal anti-inflammatory drugs on radiation-induced tissue damage. In one study, opossums were treated with 600 mg aspirin before, and twice daily for one week after exposure of the esophagus to 22.5 Gy ionizing radiation (73). This treatment regimen attenuated the radiation-induced esophagitis (73). A similar treatment regimen with indomethacin yielded the same results (26). Thus, it appears that non-steroidal anti-inflammatory drugs are radioprotective. In addition, some of the radioprotective effects of steroids may be mediated by a suppression of cyclooxygenase product synthesis.

The role of cyclooxygenase products as mediators of radiation-induced tissue injury remains controversial (26,71-73,87). On the one hand, non-steroidal anti-inflammatory drugs appear to be radioprotective (26,73,87) and on the other hand, these same drugs may have no radioprotective value (71,72). It is possible that these compounds, may have little direct radioprotective effects on cells, but may inhibit the formation of a cytotoxic agent or promote the synthesis of a cytoprotective compound in vivo (26,73). Alternatively, cyclooxygenase inhibitors, by decreasing the synthesis of highly vasoactive arachidonic acid metabolites, may either promote tissue perfusion at a time when the circulation is compromised, or induce vasoconstriction reducing oxygen delivery to the tissues and exerting an hypoxic radioprotective effect. Further studies are necessary to determine if cyclooxygenase inhibitors are radioprotectants and, if so, their mechanism(s) of action.

C. Prostaglandin analogues: A third approach to determining the role cyclooxygenase products play in radiation-induced tissue damage is to treat animals with exogenous stable prostaglandin analogues prior to irradiation (26,87-90).

Northway et al (26) treated opossums with the stable analogue 16,16 dimethyl prostaglandin E2 before, and every four hours for 24 hours after irradiation of the esophagus with 22.5 Gy. Acute esophagitis, as determined endoscopically and histologically occurred 7 to 10 days post irradiation and was more severe in the 16,16 dimethyl PGE2 treated group (26). Conversely, Hanson and Ainsworth (91) administered the same prostaglandin analogue to mice one hour before exposure to 11.0 to 18.0 Gy whole body irradiation and demonstrated a significant radioprotective effect of 16,16 dimethyl PGE2. This study confirmed an earlier investigation in which Hanson and Thomas (90) showed that mice pretreated with 16,16 dimethyl PGE2 had an increased clonogenic survival of jejunal epithelium four days after exposure. Finally Uribe et al (88) administered 16,16 dimethyl PGE2 orally (0.1 to 0.8 mg/kg) to rats 24, 8, and 0.5 hours before as well as 16 and 24 hours after irradiation. At 72 hours post irradiation, a sample of small intestine was removed 15 cm from the pyloric sphincter. The mucosal crypts of rats given low doses of 16,16 dimethyl PGE2 showed less radiation-induced damage than untreated rats or animals treated with higher doses of the same analogue (88). These studies suggest that the different responses observed with 16,16 dimethyl PGE2 may be related to the dose of analogue used.

III. SUMMARY AND CONCLUSIONS

1. Doses of radiation that cause tissue damage, are associated with an increased release of the classical prostaglandins, PGE2 and PGF2a, from tissue homogenates (32,41,42,45-48,50,89), cell cultures (92),

lung parenchymal strips (43,44), isolated perfused organs (54), and whole animals (63,64).

2. Radiation exposure is associated with alterations in TXA2 and PGI2 release by tissue homogenates (39), tissue parenchyma (43,44,51,60-62), cells in culture (56-59), isolated perfused organs (52,54), and whole animals (49,53,54,63,64).
3. Steroid pretreatment increased the radioresistance of cells (67,68,71) and reduced tissue damage to the opossum esophagus (73) but had no effect on the radiation-induced mortality (65).
4. Non-steroidal anti-inflammatory drugs increased the radioresistance (26,69,70,73,87) of tissues _in vivo_ but had no effect (71,72,90,93) on the radioresistance of cells in culture.
5. Exogenous administration of prostaglandins or stable analogues of cyclooygenase products either increased (88,90,91,94-971) decreased (26) or had no effect on (93,96) the radioresistance of various tissues and cells following _in vivo_ irradiation.

Additional studies will be required in order to elucidate these contradictory results. There are, however, several points which can be made. First, extrapolation of information from isolated organs to the whole animal should be done with caution. Second, a definitive relationship between altered cyclooxygenase product release and radiation-induced tissue damage remains to be established. If such a relationship does exist, novel treatments for the deleterious side effects of radiotherapy would be possible. Finally, an increased understanding of the molecular basis of radiation injury may contribute to an increased understanding of the synthesis and metabolism of cyclooxygenase products.

REFERENCES

1. Dusting, G.J., Moncda, S. and Vane, J.R. Eur. J. Pharmacol. _49_: 65-72, 1978.
2. Kot, P.A., Johnson, M., Ramwell, P.W. and Rose, J.C. Proc. Soc. Exp. Biol. Med. _149_: 953-957, 1975.
3. Marmo, E., Rossi, F., Lamysa, E., Giordano, L., Caputi, A.P., Vacca, C., Ariello, B. and Rosatti, F. Adv. Prostaglandin Thromb. Res. _7_: 701-709, 1980.
4. Rose, J.C., Kot, P.A., Ramwell, P.W., Doykos, M. and O'Neill, W.P. Proc. Soc. Ep. Biol. Med. _153_: 209-212, 1976.
5. Terashita, Z., Fukui, H., Nishikawa, K., Hirata, M. and Kikuchi, S. Eur. J. Pharmacol. _53_: 49-56, 1978.

6. Armstrong, J.M., Boura, A.L.A., Hamberg, M. and Samuelsson, B. Eur. J. Pharmacol. 39: 251-258, 1976.
7. Dusting, G.J., Chapple, D.J., Hughes, R., Moncada, S. and Vane, J.R. Cardiovasc. Res. 12: 720-730, 1978.
8. Fletcher, J.R. and Ramwell, P.W. Adv. Prostaglandin Thromboxane Res. 7: 723-725, 1980.
9. Rose, J.C., Johnson, M., Ramwell, P.W. and Kot, P.A. Proc. Soc. Exp. Biol. Med. 147: 652-655, 1974.
10. Charo, I.F., Feinmann, R.D., Detwiler, T.C., Smith, J.B., Ingerman, C.M. and Silver, M.J. Nature 269: 66-69, 1977.
11. Cowan, D.H. Brit. J. Haematol. 49: 425-434, 1981.
12. Hamberg, M., Svensson, J. and Samuelsson, B. Proc. Natl. Acad. Sci. U.S.A. 72: 2994-2998, 1975.
13. Hamberg, M., Svensson, J., Wakabayashi, T. and Samuelsson, B. Proc. Natl. Acad. Sci. U.S.A. 71: 345-349, 1974.
14. Malmsten, C., Hamberg, M., Svensson, J. and Samuelsson, B. Proc. Natl. Acad. Sci. U.S.A. 72: 1446-1450, 1975.
15. Adelman, B., Stumman, M.B., Mennell, D. and Handin, R.I. Blood 58: 198-205, 1981.
16. Bunting, S., Gryglewski, R., Moncada, S. and Vane, J.R. Prostaglandins 12: 897-913, 1976.
17. Gryglewski, R.J., Bunting, S., Moncada, S., Flower, R.J. and Vane, J.R. Prostaglandins 12: 685-713, 1976.
18. Higgs, E.A., Higgs, G.A., Moncada, S. and Vane, J.R. Brit. J. Pharmacol. 63: 535-539, 1978.
19. Bacq, Z.M. and Alexander, P. In: Fundamentals of Radiobiology, Pergamon Press, 1966, pp. 406-435.
20. Denekamp, J. Br. J. Cancer 36: 322-329, 1977.
21. Gross, N.J. Lung 159: 115-125, 1981.
22. Hirst, D.G., Denekamp, J. and Travis, E.L. Radiat. Res. 77: 259-275, 1979.
23. Hopewell, J.W. Radiat. Res. 63: 157-164, 1975.
24. McLennon, G., Oberley, L.W. and Autor, A.P. Radiat. Res. 84: 122-132, 1980.
25. Northway, M.G., Bennett, A., Carroll, M.A., Eastwood, G.L., Feldman, M.S., Mamel, J.J. and Szavarc, I.A. In: Prostaglandins and Cancer: First International Conference, Vol. 2 (Eds. T.J. Powles, R.S. Bockman, K.V. Honn and P. Ramwell), Alan R. Liss, Inc., New York, 1982, pp. 799-802.
26. Northway, M.G., Libshitz, H.I., Osborne, B.M., Feldman, M.S., West, J.H. and Szavarc, I.A. Gastroenterol. 78: 883-892, 1980.
27. Northway, M.G., Libshitz, H.I., West, J.H., Withers, H.R., Mukhopadhyay, A.K., Osborne, B.M., Szwarc, I.A. and Dodd, G.D. Radiobiology 131: 731-735, 1979.
28. Northway, M.G., Mamel, J.J., Libshitz, H.I., West, J.H., Feldman, M.S., Osborne, B.M., and Szavarc, I.A. Gastroenterol. 76: 1209, 1979.
29. Phillips, T.L. Seminars in Oncol. 8: 65-82, 1981.
30. Phillips, T.L., Benak, S. and Ross, G. Front. Radiat. Ther. Oncol. 6: 21-43, 1972.
31. Reinhold, M.S. Front. Radiat. Ther. Oncol. 6: 44-56, 1972.
32. Romanstev, E.F., Zhulanova, Z.I. and Nikandrova, T.I. Vestn. Akad. Med. Nauk. S.S.S.R. 9: 86-89, 1982.

33. Rosevit, B., Malsky, S.J., Reid, C.B. Am. J. Roentgenol. 114: 460-475, 1972.
34. Stearner, S.P., Christian, E.J.B. Radiat. Res. 47: 741-755, 1971.
35. Takahashi, M. and Hellman, R.F. Int. J. Radiation Oncol. Biol. Phys. 2: 61-68, 1977.
36. Harms, B.A., Bodai, B.I., Smith, M., Gunther, R., Flynn, J. and Demling, R.H. J. Surg. Res. 31: 274-280, 1981.
37. Lefer, A.M. Circ. Shock 6: 297-303, 1979.
38. Longenecker, G.L., Bowen, R.J., Eddy, I.J. and Glenn, T.M. Circ. Shock 8: 393-401, 1981.
39. Maclouf, J., Bernard, P., Rigaud, M., Rocquet, G. and Breton, J.C. Biochem. Biophys. Res. Comm. 79: 585-591, 1977.
40. Borowska, A., Sierkowski, S., Mackowiak, J. and Wisniewski, K. Experientia 35: 1368-1370, 1979.
41. Eisen, V. and Walker, D.I. Br. J. Pharmacol. 57: 527-532, 1976.
42. Nikandrova, T.I., Zhulanova, Z.I. and Romantsev, E.F. Radiobiologiia 21: 265-269, 1981.
43. Steel, L.K. and Catravas, G.N. Int. J. Radiat. Biol. 42: 517-530, 1982.
44. Steel, L.K., Sweedler, I.K. and Catravas, G.N. Radiat. Res. 94: 156-165, 1983.
45. Trocha, P.J. and Catravas, G.N. Adv. Prostaglandin Thromboxane Res. 7: 851-856, 1980.
46. Walker, D.I. and Eisen, V. Int. J. Radiat. Biol. 36: 399-407, 1979.
47. Ziboh, V.A., Mallia, C., Morhart, E. and Taylor, J.R. Proc. Soc. Exp. Biol. Med. 169: 386-391, 1982.
48. Trocha, P.J. and Catravas, G.N. Int. J. Radiat. Biol. 38: 503-511, 1980.
49. Schneidkraut, M.J., Kot, P.A., Ramwell, P.W. and Rose, J.C. Adv. Prostaglandin Thromboxane Leukotriene Res. 12: 107-112, 1983.
50. Pausescu, E., Chirvasie, R., Trodosiu, T. and Paun, C. Radiat. Res. 65: 163-171, 1976.
51. Allen, J.B., Saguman, R.H. and Stuart, M.J. Lancet 2: 1193-1196, 1981.
52. Heinz, T.R., Schneidkraut, M.J., Kot, P.A., Ramwell, P.W. and Rose, J.C. Prog. Biochem. Pharmacol. 20: 74-83, 1985.
53. Schneidkraut, M.J., Kot, P.A., Ramwell, P.W. and Rose, J.C. J. Appl. Physiol. 57: 833-838, 1984.
54. Schneidkraut, M.J., Kot, P.A., Ramwell, P.W. and Rose, J.C. J. Appl. Physiol. 61: 1264-1269, 1986.
55. Baluda, V.P., Sushkevich, N., Parshkov, E.M. and Lukoyanova, T.I. Biull. Eksp. Biol. Med. 91: 559-562, 1981.
56. Eldor, A., Vlodavsky, I., HyAm, E., Atzman, R. and Fuks, Z. Prostaglandins 25: 263-279, 1983.
57. Friedman, M., Saunders, D.S., Madden, M.C., Chaney, E.L. and Kwock, L. Radiat. Res. 106: 171-181, 1986.
58. Hahn, G.L., Menconi, M.J., Cahill, M. and Polgar, P. Prostaglandins 25: 783-791, 1983.
59. Rubin, D.B., Drab, E.A., Ts'ao, C.H., Gardner, D. and Ward, W.F. J. Appl. Physiol. 58: 592-597, 1985.
60. Sinzinger, H., Cromwell, M. and Firbas, W. Radiat. Res. 97: 533-536, 1984.

61. Sinzinger, H., Firbas, W. and Cromwell, M. Prostaglandins 24: 323-329, 1982.
62. Ts'ao, C., Ward, W.F. and Port, C.D. Radiat. Res. 96: 284-293, 1983.
63. Donlon, M., Steel, L., Helgeson, E.A., Shipp, A. and Catravas, G.N. Life Sci. 32: 2631-2639, 1983.
64. Donlon, M., Steel, L., Helgeson, E.A., Wolfe, W.W. and Catravas, G.N. Int. J. Radiat. Biol. 47: 205-212, 1985.
65. Jacob, C.O., Brenner, H. and Hercberg, A. Isr. J. Med. Sci. 20: 639-640, 1984.
66. Millar, B.C. and Jinks, S. Int. J. Radiat. Oncol. Biol. Phys. 8: 795-797, 1982.
67. Millar, B.C., Jenkins, T.C., Fielden, E.M. and Jinks, S. Radiat. Res. 96: 160-172, 1983.
68. Millar, B.C., Jinks, S. and Powles, T.J. Brit. J. Cancer 44: 733-740, 1981.
69. Mennie, A.T. and Dalley, V. Lancet 1: 1131, 1973.
70. Mennie, A.T., Dalley, V.M., Dinneen, L.C. and Collier, H.O.J. Lancet 2: 942-943, 1975.
71. Millar, B.C. Biochem. Pharmacol. 31: 2439-2445, 1982.
72. Millar, B.C. and Jinks, S. Brit. J. Radiol. 54: 505-511, 1981.
73. Northway, M.G., Eastwood, G.L., Libshitz, H.I., Feldman, M.S., Mamel, J.J. and Szwarc, I.A. Dig. Dis. Sci. 27: 923-928, 1982.
74. Bertele, V., Tomasiak, M., Falanga, A., Cerletti, C. and de Gaetano, G. Lancet 2: 775, 1982.
75. Bonney, R.J. and Humes, J.L. J. Leukocyte Biol. 35: 1-10, 1984.
76. Cerletti, C., Livio, M. and de Gaetano, G. Biochem. Biophys. Acta 714: 122-128, 1981.
77. de Kergommeaux, B.D., Ali, M. and McDonald, J.W.D. Prostaglandin Leukotriene Med. 11: 225-231, 1983.
78. Gallavan, G.H. and Jacobson, E.D. Proc. Soc. Exp. Biol. Med. 170: 391-397, 1982.
79. Gryglewski, R.J.and Vane, J.R. Mater Med. Pol. 12: 159-168, 1980.
80. Humes, J.L., Winter, C.A., Sadowski, S.T. and Kuehl, F.A. Jr. Proc. Natl. Acad. Sci. U.S.A. 78: 2053-2056, 1981.
81. Lefer, A.M. Fed. Proc. 44: 275-280, 1985.
82. Levenson, D.J., Simmons, C.E. and Brenner, B.M. Am. J. Med. 72: 354-374, 1982.
83. Miyazawa, K., Iimori, Y., Makino, M., Mikami, T. and Miyasaka, K. Jpn. J. Pharmacol. 38: 199-205, 1985.
84. Nistico, G., Rotiroti, D., GeGori, N., Pisanti, N., Mandolfino, T. and Tommasini, R. Res. Comm. Chem. Pathol. Pharmacol. 32: 245-249, 1981.
85. Roth, G.J., Stanford, N. and Majerus, P.W. Proc. Natl. Acad. Sci. U.S.A. 72: 3073-3076, 1975.
86. Vane, J.R., Bunting, S. and Moncada, S. Int. Rev. Exp. Pathol. 23: 161-208, 1982.
87. Ambrus, J.L., Ambrus, C.M., Lillie, D.B., Johnson, R.J., Gastpar, H. and Kishel, S. J. Med 15: 81-92, 1984.
88. Uribe, A., Johansson, C., Rubio, C. and Arndt, J. Acta Radiol. Oncol. 23: 349-352, 1984.
89. Trocha, P.J. and Catravas, G.N. Radiat Res. 94: 239-251, 1983.
90. Hanson, W.R. and Thomas, C. Radiat. Res. 96: 393-398, 1983.
91. Hanson, W.R. and Ainsworth, E.J. Radiat. Res. 103: 196-203, 1985.

92. Gibson, D.P., DeGowin, R.L. and Knapp, S.A. Radiat. Res. 89: 537-545, 1982.
93. Millar, B.C. and Jinks, S. Int. J. Radiat. Biol. 46: 367-373, 1984.
94. Tomas-de la Vega, J.E., Banner, B.F., Hubbard, M., Boston, D.L., Thomas, C.W., Straus, A.K. and Roseman, D.L. Surg. Gynecol. Obstet. 158: 39-45, 1984.
95. Chirkov, Y.Y. and Sobolev, A.S. Strahlentherapie 160: 521-522, 1984.
96. Das, U.N., Ramadevi, G., Rao, K.P. and Rao, M.S. Prostaglandins 29: 911-920, 1985.
97. Prasad, K.N. Int. J. Radiat. Biol. 22: 187-189, 1972.

Supported by: U.S. Army Contract DAMD17-84-C-4006 and
ONR grant N00016-82-K-0713

4

IRRADIATION AND BLOOD VESSEL PROSTAGLANDIN METABOLISM

H. SINZINGER and R. REICHEL

Dept. of Nuclear Medicine and Atherosclerosis Research Group (ASF), Vienna (H.S.),
1st Dept. of Gynecology and Obstetrics (R.R.), University of Vienna, Austria

INTRODUCTION

Nowadays, it is well known, that tissue damage is associated with thrombosis of vessels in irradiated areas. As early as in 1944 Sheehan (43) was the first relating the foam cell plaque formation to an effect of irradiation. Warren (61) reported in 1942 atherosclerosis-like lesions. Later on, a number of experimental studies have been performed in order to elucidate the pathomechanism of irradiation- induced vasculopathy showing in rat (2), chicken (58) and swine (30) that irradiation accelerates the development of atherosclerosis. Endothelial damage (9, 10, 25, 27, 28, 29, 57), increased vascular wall permeability (35), enhanced mitotic activity (14, 44, 46) and parietal thrombus formation (6, 15, 16) have been shown to be early mechanisms involved. Human data provide evidence for cerebrovascular consequences (7) and the development of coronary heart disease induced by irradiation (5, 38). Although all these mechanisms together with the finding, that aspirin reduces the severity of radiation-induced vascular lesions (56) indicate, that the prostaglandin (PG)-system is involved to an important extent too, there is only limited information available so far concerning the importance of the PG-system in the blood- vessel wall interaction during and after irradiation.

MATERIAL AND METHODS

Irradiation technique in rabbits.

One-year old male rabbits were used for our studies. The irradiation- technique with Co-60 γ-rays using a focus-skin distance of 40 cm has been described in detail earlier (44-46). The animals were irradiated under general anaesthesia (Aethyl-(1-methylbutyl)- thiobarbituric acid sodium, Sanabo, Vienna, ip) with a single or repeated dose of 1, 5 or 10 Gy to a 10 x 10 cm area of the abdomen (including the whole abdominal aorta). In another group of animals the dose of 5 Gy was applied again one week later. Control animals were sham treated. Different time intervalls after irradiation, the rabbits were painlessly killed by a blow on their neck. 30 minutes before killing they received 100 IU heparin iv in order to prevent perimortal thrombin-induced PGI2-formation in-vitro. Immediate laparotomy was performed and the irradiated segment of the abdominal aorta as well as a non-irradiated thoracic aortic control segment were prepared in an atraumatic way in ice-cold buffer (tris-HCl, pH 7.4). The excised abdominal aortic segments were immediately frozen and stored at

P. Polgar (ed.), EICOSANOIDS AND RADIATION.

-70° C until estimation of biosynthetic capacity of PGI2 in-vitro. In addition, in some of the animals a small segment was used for radiothinlayer-chromatographical studies. Prostacyclin (PGI2) formation was assessed using the

platelet aggregation (inhibition) bioassay (34, 47):

human venous blood was anticoagulated using 3.8 % sodium citrate (1:9). After a 10 minutes sedimentation at 22° C platelet-rich plasma (PRP) was obtained by centrifugation at 150 g for 5 minutes. After the removal of the PRP a platelet-poor plasma (PPP) was obtained using a further centrifugation step at 1000 g for 10 minutes. PRP was adjusted with PPP to a final platelet count of 250 x 10^3/µl. After prewarming of the PRP samples at 37°C in a Born-type aggregometer and addition of 100 µl tris-HCl buffer (pH 7.4), the aggregation was induced by 1-µM ADP (Sigma) dissolved in 100 µl. The aggregation response was quantified using the tangent (α) of the angle of the aggregation curve and the maximal amplitude (%).

Irradiation technique in women with cervical cancer (36):

11 women, aged 40 to 76 years (59 ± 10.2) with squamous cell carcinoma of the uterine cervix, 8 of them in clinical stage II and 3 in clinical stage III (FIGO classification, 21) underwent irradiation therapy with a total local dose of 60 Gy. Before, immediatly after, 8 and 20 weeks after irradiation therapy blood samples were taken for determination of TXB2 in plasma and serum, platelet aggregation, testing of platelet sensitivity to antiaggregatory prostaglandins and of the conversion rate of exogenously radiolabelled arachidonic acid (AA) to various eicosanoids by means of radiothinlayer-chromatography(RTLC).

Radiothinlayer - chromatography (RTLC).

Human blood was anticoagulated using 3 volumes of acid citrate dextrose (ACD) and 7 volumes of blood. Only synthetic material or siliconized glass pipettes were used. After a 15 minutes sedimentation at room temperature (22°C), the samples were centrifuged for 7 minutes at 150g (22°C) without using the brake. From the PRP 500 µl were used after centrifugation. A platelet pellet is obtained by 5 minutes centrifugation at 700g at room temperature. The pellet is resuspended in Tris-HCl-buffer. 0.25 µCi ^{14}C-AA (59,6 µCi/mM, Amersham, UK) is added and this mixture is incubated in a shaking water bath at 37°C for 5 minutes. The reaction is stopped using 1N-HCl at pH 3. After centrifugation for 10 minutes at 5000g and 4°C the supernatant is removed and transfered into another syringe. 2 ml ethylacetate were added and mixed with a vortex for 3 minutes. After 60 minutes incubation at 4°C and a further centrifugation step for 10 minutes at 4°C and 5000g 1.5ml of the upper organic phase are removed and transfered into conic syringes. 1 ml ethylacetate was added again and mixed using a vortex. After centrifugation under the same conditions 1 ml of the upper organic phase is removed and added into the conic syringes. The samples are dried under nitrogen dissolved in 100 µl absolute ethanol and stored at -20°C. 15µl of the samples are added to a thinlayer plate, developed in a chamber, dried and measured using a TLC-linear analyser (Berthold Comp.).

The conversion rate of ^{14}C-AA was expressed in percent (%).

Radioimmunoassay for TXB2 in plasma.

Venous blood (1:9) was anticoagulated with 2% sodium-2-EDTA and 10 mg Aspisol for cyclooxygenase inhibition. After a 30-minutes rest blood was drawn without venous occlusion at 4° C. After a short sedimentation (22° C, 10 minutes) the samples have been centrifuged at 1500 g for 15 minutes. The carefully separated plasma was stored at -70° C and estimated for TXB2 within 2 weeks in order to avoid artificial increase (54). Serum TXB2 formation: 2 ml of whole blood were allowed to clot at 37° C in a glass syringe for 60 minutes. Thereafter, the samples have been centrifuged after clot removal at 4° C at 1500 g for 15 minutes. The samples were stored at -70° C until the investigation by radioimmunoassay, again being performed not later than 2 weeks afterwards.

Radioimmunoassay (RIA)

RIA is performed using non-extracted samples and the double-antibody- technique. The samples are mixed with normal rabbit plasma ^{3}H-TXB2 (128,0 Ci/mM, NEN, Dreieich, FRG; 4000 cpm per 0.5 ml), 2.5 ml à 0.01M Gel-tris-buffer (pH 7.4, 1 g gelatine/l) and 100 µl of a specific antibody. After mixing, the ingredients are incubated in a shaking water bath at 37°C for 1 hour. For the separation of free and bound antigen 0.1 ml antibody against rabbit gamma-globuline from the goat (Behring, Marburg, FRG) are added and followed by an overnight incubation at 4°C. The antigen-antibody-complex is centrifuged on the next day at 4°C and 1500 g. The supernatant is removed and the precipitate washed using 0.5 ice-cold 0.9% NaCl. Thereafter, the samples are centrifuged at 4°C and 1500g. Samples are dissolved in 10 ml Picofluor 30 (Packard, Warrenville, Illinois, USA) and the radioactivity is measured in a gamma-counter for 10 minutes. The interassay variation amounts 6.5%, the intraassay variation 4.3%.

Platelet sensitivity testing.

For the determination of platelet sensitivity instead of 100 µl buffer samples at least 3 different concentrations of either PGE1 or PGI2 (the standards have been kindly supplied by Dr. John E. Pike, the Upjohn Company, Kalamazoo, Michigan, USA) were added dissolved in the same volume to the PRP 1 minute prior to induction of aggregation by means of ADP. The platelet sensitivity was calculated (48) as the amount of PG necessary to suppress the induced aggregation to the half (ED 50). This amount calculated from different experiments has been verified by final titration. The higher the amount of PG necessary to lower the aggregation response the lower the sensitivity of the platelets to this particular PG.

RESULTS

Radiothinlayer chromatography of the normal rabbit vascular tissue shows that PGI2 (as measured by its stable derivate 6-oxo-PGF1α) is the main metabolite of AA. Both single and repeated irradiation demonstrate that in the early phase an increase (figure 1, table 1) in the conversion from exogenously added AA occurs, whereas later on, a significant drop in the conversion rate can be seen.

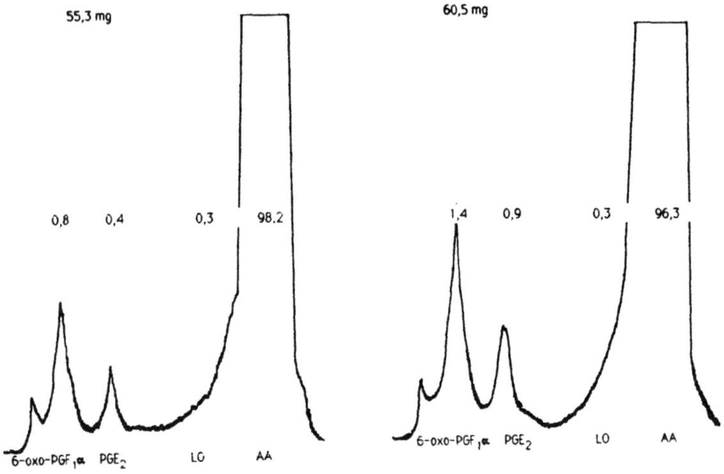

Fig. 1. Radiochromatogram of rabbit abdominal aorta before and 2 hours after irradiation. There is a clear increase in the cyclooxygenase products (6-oxo-PGF1α and PGE2, respectively), whereas the lipoxygenase (LO) is unchanged.

Table 1. RTLC of irradiated rabbit aorta before and after 10 Gy-irradiation, Mean ± S.D.; n = 4; *) $p \leq 0,01$; conversion rate in %.

time (hours)	6-oxo-PGF1α	PGE2	LO-products
Co	4,8 ± 1,2	2,6 ± 0,6	2,9 ± 0,7
2	7,3 ± 2,6*	3,9 ± 0,9*	3,4 ± 0,6
4	6,8 ± 2,4*	2,7 ± 0,8	3,1 ± 0,9
8	5,1 ± 1,9	2,1 ± 0,6	2,9 ± 0,8
24	4,5 ± 1,0	2,4 ± 0,7	3,0 ± 0,5
48	3,6 ± 0,9*	1,7 ± 0,5*	2,6 ± 0,4

Only minor amounts of other compounds, and especially no thromboxane formation can be noted. No significant change in any other of the arachidonic acid-conversion products occured. These findings are paralleled by a dose dependent increase (figure 2) in PGI2-formation as determined by means of the platelet aggregation bioassay (figure 3). Repeated irradiation causes a further temporary increase in PGI2-production. This initial increase is followed by a subsequent long-lasting decrease in

PGI2-formation (figures 2, 4) persisting for the entire observation period of up to 12 months. The higher the radiation dose the more pronounced the drop and the more severe the depression in PG-generation lateron.

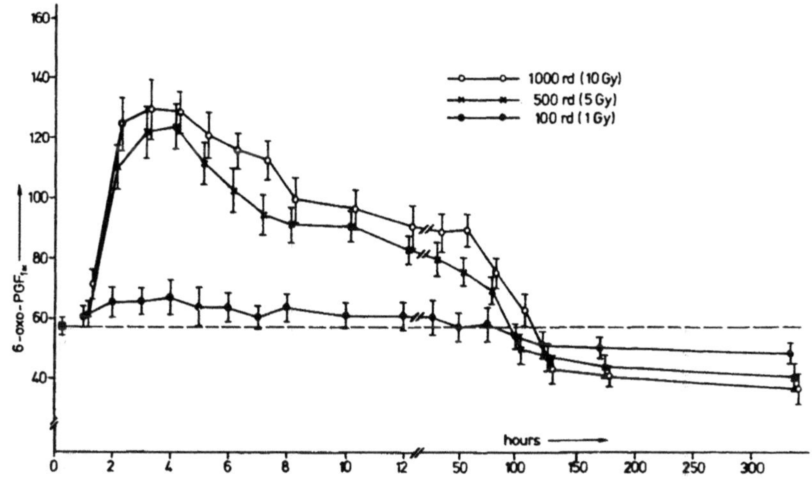

Fig. 2. PGI2-formation (as assessed radioimmunologically by its stable derivate 6-oxo-PGF1α) after various radiation doses and time intervalls. After a dose-dependent, short-lasting increase, a persistent, long-lasting diminution in PGI2-formation can be seen. The straight line reflects PGI2-generation of control vascular segments (mean value).

Radioimmunological determination of the vascular wall PGI2-synthesis (assayed by means of specific RIA for 6-oxo-PGF1α) reveals a comparable behaviour to the bioassay data. Whereas control vascular segments are able to produce significantly increased amounts of PGI2 in vitro upon PDGF-stimulation dose dependently, no such response can be seen in irradiated segments (figure 5). There is, however, a trend for a decreased response after higher radiation dose exposure. However, using a specific TXB2-assay minimal amounts of this substance can be measured showing an acute increase due to irradiation persisting for months above the control values (figure 6). Preliminary findings from irradiation in man suggest that a moderately increased in-vivo TX-release (reflected by plasma TXB2) from platelets (or, however, unlikely, in part from the vessel wall) is paralleled by a decreased platelet TX-formation stimulated by endogenous thrombin in-vitro during clotting (table 2).

During irradiation therapy of women suffering from cervical cancer plasma TXB2 exhibits a continuous increase throughout the observation period. In parallel, there is a marked decrease in serum TXB2 at the same time. The peripheral platelet count does not show any change. The plasma 6-oxo-PGF1α is unaltered too (table 2).

Table 2. Metabolites of the prostaglandin pathway in plasma (pg/ml) and in serum (ng/ml) and platelet count (x1000/µl) before (0), immediately after (1), 8 weeks (2) and 20 weeks (3) after irradiation therapy of cervical cancer, Mean ± S.E.

		0	1	2	3
TXB2	plasma	7,8 ± 3,9	9,6 ± 4,6	13,0 ± 4,8	15,4 ± 4,3
	serum	13,2 ± 6,6	0,6 ± 0,3	1,1 ± 0,4	0,6 ± 0,2
6-oxo-PGF1a	plasma	2,4 ± 1,3	4,4 ± 2,6	7,0 ± 5,8	7,7 ± 3,1
plateletcount		320,0 ±40,2	314,0 ±39,2	350,0 ±44,0	297,0 ±34,2

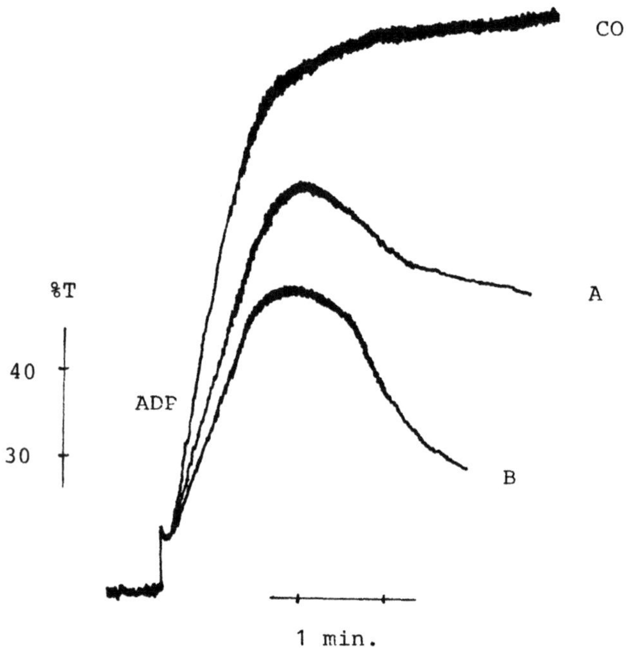

Fig. 3. Platelet aggregation bioassay for the measurement of biologically active PGI2. The incubation buffer suppresses the 1 µM ADP-induced aggregation dependent on the amount of biologically active PGI. CO...control curve, A...4 hours after irradiation, B...prevalue (5 mg tissue wet weight).

The platelet sensitivity behaviour (table 3) demonstrates a continuous decrease in the amount of antiaggregatory PG necessary to inhibit the ADP-induced aggregation response being indicative for an increase in platelet sensitivity for both the PG-compounds (PGI2, PGE1) examined. The RTLC (table 4, figure 7) shows a decrease in the conversion of exogenously added AA to the cyclooxygenase compounds, namely TXB2 and HHT, resulting in a shift to the lipoxygenase (LO)-pathway, i.e. 12-HETE.

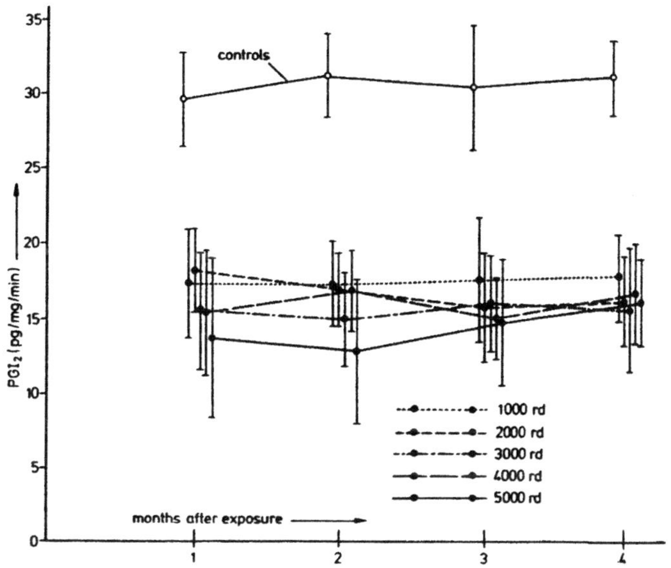

Fig. 4. Months after a single dose irradiation a significantly depressed ($p<0,01$) vascular PGI2-formation still can be monitored. Although no significant differences exist, there seems to be a trend for a lower PGI2-synthesis the higher the radiation dose was.

Table 3. Platelet sensitivity to PGI2 and PGE1, Mean ± S.E., (ng/ml PRP)

	0	1	2	3
PGI2	2,6 ± 0,3	2,2 ± 0,4	2,5 ± 0,4	1,5 ±0,3
PGE1	9,7 ± 1,4	10,1 ± 1,8	10,1 ± 2,2	7,7 ± 1,3

Table 4. Metabolites of arachidonic acid (AA) from human platelets during and after irradiation therapy of cervical cancer, RTLC, Mean ± S.E., conversion rate in %.

	0	1	2	3
AA	6,6 ± 1,2	7,0 ± 1,4	11,2 ± 6,7	13,0 ± 3,2
TXB2	23,9 ± 3,2	24,5 ± 4,0	21,1 ± 4,2	3,8 ± 3,8
HHT	26,2 ± 2,9	19,9 ± 3,0	18,5 ± 5,6	17,5 ± 3,1
12-HETE	39,0 ± 3,9	41,2 ± 6,4	43,4 ± 7,3	49,4 ± 6,4

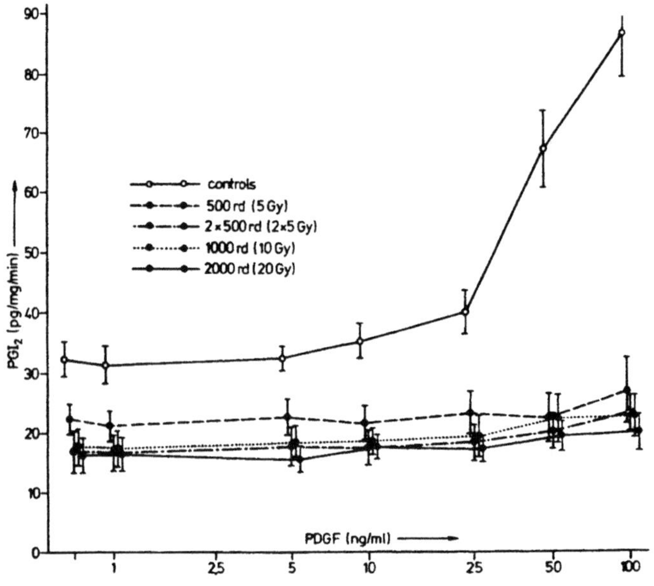

Fig. 5. Response of vascular wall PGI2-synthesis upon PDGF-stimulation. Whereas vascular control segments exhibit a significantly enhanced PGI2-generation, nearly no response occurs in irradiatedaorticsegments.

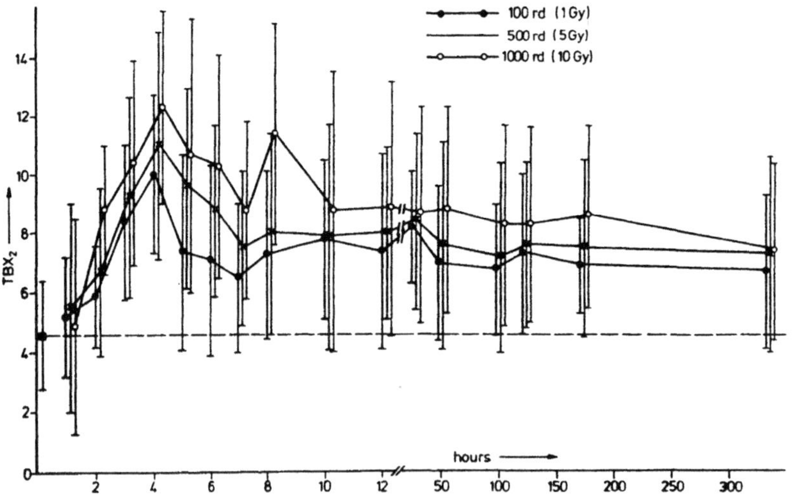

Fig. 6. TXB2 values reflecting TXA2-synthesis by the arterial wall cells. Same experiments as in figure 2. After a peak formation about 6 hours after irradiation, a persistent and long-lasting enhanced production can be monitored. No significant dose-dependency can be discovered.

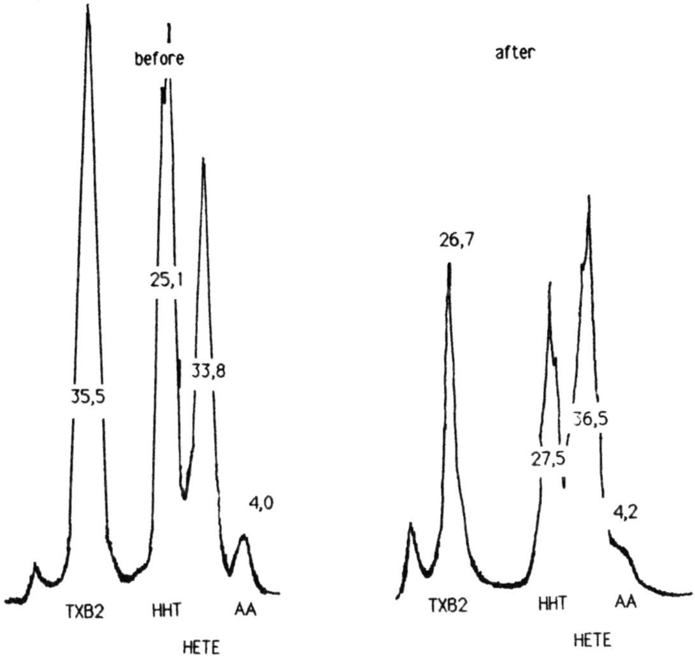

Fig. 7. Typical chromatograms. RTL-chromatogram of the platelets of a 32 years aged female (M.S.) suffering from cancer of the uterine cervix before and 4 weeks after irradiation. The shift in the conversion of exogenous ^{14}C-AA from TXB2 to lipoxygenase products is apparent.

DISCUSSION

Tissue damage and vascular thrombosis (14-16) as well as atherosclerosis-like lesions in experimental conditions (31, 44) and in man (45, 61, 63) are well known complications of irradiation. Most likely, the early endothelial damage (9, 32, 35, 37) is paralleled by an increased PGI2-generation simply due to enhanced liberation of endogenous precursor fatty acids (18) and an activation of cyclooxygenase (22). This early severe increase in the antiaggregatory PGI2 (34) which has also been monitored dose dependently in cultured endothelial cells (22, 40) may cause on the one hand vasodilatation and an increased permeability (35) and may furthermore inhibit platelets at a local level to form mural vascular thrombi. The decreased PGI2 - formation in later stages after irradiation (3, 4, 49, 51, 52, 55) correlates well with Allen's findings (1) of a diminished activity of PGI2- synthetase in endothelial lining. Eldor (11, 12) observed a decreased PGI2-forming capacity by endothelial cells surviving irradiation, whereas Weksler (60) found PGI2 being only slightly affected. Menconi (33) reported an effect on endothelial cells only, indicating an unique sensitivity of this cells, whereas smooth muscle cells and fibroblasts did not respond similarily. However, the mechanism of the finding, either a free radical fatty acid or peroxide-induced enzyme inhibition is not yet clarified. For the long-lasting effects the interaction of prostacyclin and the platelet derived growth factor (PDGF), a mitogen being derived from activated platelets, in the vessel wall might be of key importance. On the one hand the PDGF-liberation (39) from the platelets is enhanced during their activation as occuring during irradiation (figure 8). On the other hand, in turn, the PGI2 formed by the vessel wall, especially in the early phase, is able to inhibit selectively PDGF-release from the platelets. Subsequently, its proliferative and mitotic activity, which is known to be enhanced (14, 15, 44, 46), might be backregulated. Therefore, the capacity of the vessel wall to respond onto PDGF-liberation with an enhanced PGI2- synthesis (8, 50) will be one of the key mechanisms in pathogenesis of the progression of radiation-induced vasculopathy (53).

From basics it is, therefore, easy to understand, that the presence of elevated blood lipids and in particular of cholesterol would favour the foam cell formation (43) especially insofar as decreased local PGI2-levels are favouring intracellular cholesterol-synthesis and decrease the intracellular degradation (23). Furthermore, malondialdehyde (figure 8) derived from activated platelets together with LDL resulting in malonyl-LDL (17) formation may increase the cellular entry of cholesterol via a specific receptor finally resulting in foam cell formation which has been seen for the first time in irradiated human arteries by Sheehan (43).

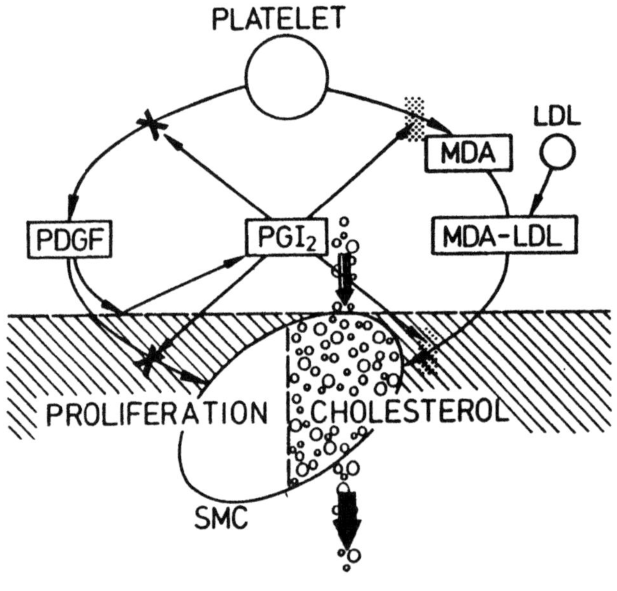

Fig. 8. Hypothetic interaction between the platelet and the vessel wall during irradiation. PDGF and MDA on the one hand and PGI2 on the other are the main hemostatic balancing factors.

As it has been shown that both the antiaggregatory prostaglandins E1 and I2 (62) as well are able to decrease mitotic activity in experimental animals and they exert an antiproliferative action on vascular smooth muscle cells in man (as well, it is under investigation at the moment whether these compounds might be of preventive effect in radiation-induced vasculopathy. Furthermore, the assumption that PGI2 acts in this context more via stabilisation of platelet membrane selectively decreasing PDGF-liberation whereas the PGE1 has a more direct antiproliferative activity on the cell itself offers an interesting idea for a concomitant paralleled treatment during irradiation in order to minimize tissue damage in the non-tumor area. Gerrard (20) recently overviewed the literature by summarizing that so far, no changes in platelet TXA2 have been reported. After renal irradiation in rat (13) found an increase in TXB2 in venous and arterial effluent, whereas 6-oxo-PGF1α in the venous part did not change. Irradiation of rat lung has been shown to be accompanied by a temporarily decreased PGI2-formation, followed later on by a long-lasting increase (59).

Concerning the platelet function even much less information exists. A few other spot-light data like a decrease in platelet sensitivity after local irradiation has been described (26); however, this finding still lacks verification at the receptor level. Furthermore, an enhanced degradation of PGI2 in irradiation has been claimed (19). Data of an in-vivo platelet activation in man as well as a - most probably pathophysiologically unimportant increased TXA2 formation by the vascular wall - is rather nicely reflected by a shift towards a relative increase in whole body TX-production after whole body irradiation in mice (24, 41, 42). This imbalance between PGI2 and TXA2 has been claimed as a key mechanism for late radiation effects (Eldor 13). The increased plasma TXB2 monitored by us, indicating an in-vivo platelet activation, does not fit together with a severely decreased serum TXB2-formation, as the actual amounts differ by about 5 orders of magnitude. In addition, a chemotactic activity attracting ventraphils, a liquid generated by peroxidation of membrane phospholipids (found to be produced by cultured vascular endothelial cells) may account for the response observed following ionizing irradiation (11).

Summarizing, the hyperactivation of platelets and the response of the irradiation-damaged vessel wall are thus the key regulators for the genesis of vasculopathy. This hypothesis needs clinical proof by a concomitant antiplatelet therapy,however, an approach which surprisingly has not been done until now.

Acknowledgements: The authors gratefully acknowledge the valuable help of Claudia Dadak, Sonja Reiter, Eva Unger, M. Cromwell, K. Ettl, W. Firbas, K. Silberbauer and R. Willvonseder.

REFERENCES

1. Allen, S.B., Sagerman, R.H. and Stuart, M.J. Lancet ii: 1193-1196, 1981.
2. Amromin, G.D., Gildenhorn, H.L., Soloman, R.D., Nodkarni, B.B. and Jacobi, M.L. J. Atheroscler. Res. 4: 325-334, 1964.
3. Baluda, V.P., Sushkevich, G.N., Parshkov, E.M. and Lukoianova, T.I. Bull. Eksp. Biol. Med. 91: 559-562, 1981.

4. Baluda, V.P., Sushekwich, G.N., Parshkow, E.M. and Lukoyanova, T.I. Zakbla Medizinski. 62: 243-247, 1982.
5. Brosius, F.C., Woller, B.F. and Roberts, W.C. Amer. J. Med. 70: 519-530, 1981.
6. Brown, J.M., Fajardo, L.F. and Stewart, R.J. Arch.Pathol. 96: 1-8, 1973.
7. Conomy, J.P. and Kellermeyer, R.W. Cancer 36: 1702-1708, 1975.
8. Coughlin, S., Moshkowitz, M.A., Zellev, P.R., Antoniades, H.N. and Levine, G. Nature 188: 5791-5794, 1980.
9. DeBoer, W.G.R.M. thesis, Univ. Groningen, 1963.
10. DeGowin, R.L., Lewis, J.L., Hoak, J.C., Mueller, A.L. and Gibson, D.P. J. Lab. Clin. Med. 84: 42-48, 1974.
11. Eldor, A., Vlodavsky, I., Hy Am, E., Atzmon, R. and Fuks, Z. Prostaglandins 25: 263-279, 1983.
12. Eldor, A.: pers. communication, 1986.
13. Eldor, A., Cohen, M., Vlodavsky, I., Pazin, E., Hy Am, E., Fuks, Z. and Matzner, Y. Int. Conf. Prostaglandins and Lipid Metabolism in Radiation Injury, Washington 1986, "Abstract book", pp. 83.
14. Fajardo, L.F. and Stewart, L.R. Amer. J. Pathol. 59: 299-306, 1970.
15. Fajardo, L.F. and Brown, M.J. Rad. Res., 55: 387-389, 1973.
16. Fajardo, L.F. and Stewart, L.R. Lab. Invest. 29: 244-257, 1973.
17. Fogelman, A.M., Shehter, I., Slager, J., Hokam, M., Child, J.S., Edwards, P.A. Proc. Nat. Acad. Sci. USA 77: 2214-2218, 1980.
18. Friedman, M., Saunders, D.S., Madden, M.C., Chaney, E.L. and Kwock, L. Radiat. Res. 106: 171-181, 1986.
19. Gabrielian, E.S., Akopov, S.E., Kharatian, A.A. and Petrosian, A.K. Biull. Eksp. Biol. Med. 101: 79-81, 1986.
20. Gerrard, J.M. In: Prostaglandins and Leukotrienes. Blood and Vascular Cell Function. Marcel Dekker Inc., New York, Basel, 1985, pp. 222.
21. Gitsch E. In: Gynäkologie, (Eds. Gitsch E, Janisch H.) Wilhelm Maudrich, Wien-München-Bern, 1984, p. 103.
22. Hahn, G.L., Menconi, M.J., Cahill, M. and Polgar, P. Prostaglandins 25: 783-791, 1983.
23. Hajjar, D.P. Biochem. Pharmacol. 34: 295-300, 1985.
24. Heinz, T.R., Schneidkraut, M.Y., Kot, P.A., Ramwell, P.W. and Rose, J.C. Biochem. Pharmacol. 20: 74-83, 1985.
25. Hirst, D.G., Denekamp, J. and Travis, E.L. Radiat. Res. 77: 259-275, 1979.
26. Kieback, D.G., Briel, R.R., Schneider, A.I., Schmucker, R.G.A. and Lippert, T.H. Thromb. Haemost. 50: 455, 1983.
27. Lamberts, H.B. In: Symposium on the cellular basis and etiology of late somatic effects of ionizing radiation, (Eds. H. Lamberts), Academic press, New York, 1963, pp. 106-125.
28. Law, M.P. and Thomlinson, R.H. Brit. J. Radiol. 51: 845-904, 1978.
29. Law, M.P. Adv. Radiation Biol. 9: 37-93, 1981.
30. Lee, K.T., Jarmolych, S., Kim, D.N., Grant, C., Krasney, J.A., Thomas, W.A. and Bruno, A.M. Exp. Molec. Pathol. 15: 170-190, 1971.
31. Lindsay, S., Kohn, H.I., Dakin, R. and Jew, J. Circ. Res. 10: 51-56, 1962.
32. Marcial-Rojas, R.A. and Castro, J.R. Ann. ORL. 71: 745-750, 1962.
33. Menconi, M.J., Taylor, L. and Polgar, P. Int. Conf. Prostaglandins and Lipid Metabolism in Radiation Injury, Washington 1986, "Abstract book", pp. 19.
34. Moncada, S., Gryglewski, R., Bunting, S. and Vane, J.R. Nature 263: 663-665, 1976.
35. Neumayr, A. and Thurnher, B. Strahlentherapie 84: 297-302, 1951.
36. Reichel, R., Husslein, P., Kucera, H. and Sinzinger, H. In: Prostaglandins in Clinic and Research, (Eds. Sinzinger, H. and Schrör, K.), Alan R. Liss, New York, 1987, in press.
37. Reinhold, H.S. and Buisman, G.H. Brit. J. Radiol. 46: 54-57, 1973.
38. Mc Reynolds, R.A., Gold, G.L. and Roberts, W.C. Amer. J. Med. 60: 39-45, 1976.
39. Ross, R., Glomset, J., Kariya, B. and Harker, L. Proc. US Nat. Acad. Sci. 74: 1207-1210, 1974.
40. Rubin, D.B., Drab, E.A., Tsao, C.H., Gardner, D. and Ward, W.F. J. Appl. Physiol. 58: 592-597, 1985.

41. Schneidkraut, M.J., Kot, P.A. and Ramwell, P.W. Adv. PG TX LT Res. 12:107, 1983.
42. Schneidkraut, M.J., Kot, P.A. and Ramwell, P.W. and Rose, J.C. J. Appl. Physiol. 57: 833-838, 1984.
43. Sheehan, J.F. Arch. Pathol. 37: 297-303, 1944.
44. Silberbauer, K., Willvonseder, R., Feigl, W., Krisch, K. and Sinzinger, H. VASA 5: 215-223, 1976.
45. Silberbauer, K., Willvonseder, R., Feigl, W. Leithner, Ch. and Sinzinger, H. VASA 6: 183-187, 1977.
46. Sinzinger, H., Wicke, L., Willvonseder, R., Hernuss, P., Müller-Tyl, E. and Erd, W. Acta Morphol. Neerl. Scand. 14: 201-208, 1976.
47. Sinzinger, H., Silberbauer, K. and Auerswald, W. Blood Vessels 17: 58-60, 1980.
48. Sinzinger, H., Schernthaner, G. and Kaliman, J. Prostaglandins 22: 773-781, 1981.
49. Sinzinger, H., Firbas, W. and Cromwell, M. Prostaglandins 24: 323-329, 1982.
50. Sinzinger, H., Kefalides, A. and Hoche, Ch. Circulation 66: 192, 1982.
51. Sinzinger, H., Firbas, W. and Cromwell, M. Wr. Klin. Wschr. 21: 761-765, 1983.
52. Sinzinger, H., Firbas, W. and Cromwell, M. Rad. Res. 97: 533-536, 1984.
53. Sinzinger, H. and Firbas, W. Brit. J. Radiol. 58: 1023-1026, 1985.
54. Sinzinger, H., Reiter, Sonja and Peskar, B.A. In: Prostaglandins and other Eicosanoids in the Cardiovascular System, (Eds. Schrör K.), Karger Verlag, Basel, 1985, pp. 62-67.
55. Sinzinger, H., Cromwell, M. and Firbas, W. Strahlentherapie und Onkologie 162: 592-595, 1986
56. Smit Sibinga, C.T. Bibl. Anat. 12: 242-248, 1973.
57. Smith, C. and Loewenthal, A. Proc. Soc. Exp. Biol. Med. 75: 859, 1950.
58. Stearner, S.P. and Sanderson, M.H. Rad. Res. 49: 328-336, 1972.
59. Tsao, C.H., Ward, W.F. and Port, C.D. 96: 284-293, 1983.
60. Weksler, Z., Raz, A., Rosenman, E., Biran, S., Fuks, Z. and Eldor, A. Int. Conf. Prostaglandins and Lipid Metabolism in Radiation Injury, Washington, 1986, "Abstract book", pp. 85.
61. Warren, S. Arch. Path. 34: 1070-1076, 1942.
62. Willis, A.L., Smith, D.L., Vigo, C. and Kluge, A.F. Lancet ii: 682-683, 1986.
63. Zollinger, H. Pathol. Europ. 5: 45-57, 1970.

5

EFFECT OF ULTRAVIOLET RADIATION ON EICOSANOID METABOLISM OF CELLS IN CULTURE

V. DELEO

Department of Dermatology, Columbia University, 630 West 168th Street, New York, NY 10032, USA

INTRODUCTION

Solar ultraviolet radiation (UV) is a ubiquitous environmental agent known to produce numerous effects in human skin. These effects include inflammation, pigmentation, carcinogenesis, local and systemic immune suppression, aging and Vitamin D metabolism. In addition, such radiation can produce cutaneous disease in individuals rendered abnormally photosensitive by the presence of exogenous drugs and chemicals or endogenous substances such as porphyrins. The mechanisms by which such radiation evokes most of these cutaneous photobiological effects have been studied in some detail but the responses remain poorly understood.

The short term response of the skin to solar UV is the classic "sunburn" response. Analogous to other wounding responses, sunburn is composed of inflammatory and repair phases. Mechanistic studies in intact skin suggest that the inflammatory response involves the production of various metabolites of arachidonic acid (AA) (1). (This is discussed in detail by H. Lim elsewhere in this volume.) The repair phase of the response is signaled by increased DNA and protein synthesis, after an initial inhibition, in irradiated skin (2). There is evidence to suggest that AA metabolites are also involved in inducing the repair phase of the response (3).

Ultraviolet radiation has long been known to be a complete carcinogen; it acts at the initiation stage of tumor induction through photon-DNA interaction. It is possible that such radiation may also play a role at the promotion stage of tumorogenesis (4-6). Chemical tumor promoters are thought to mediate promotion through "phospholipid signalling" and AA metabolite production (7). By analogy it is possible that UV effects on cellular membrane and AA metabolism likewise play an important role in cutaneous carcinogenesis.

P. Polgar (ed.), EICOSANOIDS AND RADIATION.

In addition to sunburn and carcinogenesis, UV-induced pigmentation and alteration of immune function have also been linked to UV induced changes in AA metabolism (8,9).

The skin is a complex organ composed of a large number of different cell types. Although various AA metabolites have been recovered from human and animal skin irradiated with solar radiation and radiation from artificial sources, the cell of origin of these metabolites and the mechanism of UV-induced production is unknown. In order to more clearly define these parameters of the UV-AA relationship, we and others have turned to various cells in culture as model systems. This communication is the summary of that work.

CELL CULTURE SYSTEMS

C3HIOT1/2 Cells.

In our initial studies of the effect of UV on mammalian cells we examined the effect of UVB (290-320 nm) on a marine fibroblast line - C3HIOT1/2 (10). In the standard experiment the cells were grown in dishes and prelabelled with [^{3}H] AA and irradiated with a florescent tube source (continuous spectrum in the UVB range). After irradiation the cells were reincubated and at selected times post irradiation they were assayed for release of label into culture medium. This is a measure of the UV induced release of all labelled AA metabolites. As shown in Table I, UVB stimulated release of label in a dose dependent manner for doses of radiation between 600 and 1200 J/m^2 when assayed 4 hours post irradiation.

Table 1. The effect of UVB on release of label from C3H1OT1/2 cells prelabelled with [^{3}H] arachidonic acid.

UVB dose J/m^2	Mean cpm X 10^{-3} per dish
0	14
200	18
400	20
600	25
800	46
1000	96
1200	120

Time course studies revealed that the UVB induced label release was biphasic with the initial response occurring within the first 15 minutes of irradiation. The second phase was a more gradual increase occurring between 1 and 4 hours post irradiation.

These findings are significant in biologic terms. A dose of 300-600 J/m^2 from the light source used in these studies is sufficient to induce minimal erythema, "sunburn," in human skin (minimal erythema dose - MED).

Media from irradiated and control cells were further examined to identify specific [^3H] AA metabolites produced. Media were extracted and extracts were examined by TLC with known standards. In those studies UVB stimulated the production of prostaglandins (PG) E_2, $F_{2\alpha}$ and D_2 and thromboxane (TBX) and the prostacyclin derivative 6-keto-$PGF_{1\alpha}$.

Mechanistic studies in this model indicated that the UVB stimulation of [^3H] AA release was a calcium dependent phenomenon and could be inhibited by two known phospholipase A_2 inhibitors, dexamethasone and p-bromophenacylbromide.

It is usually accepted that the first and rate-limiting step in the generation of AA metabolites is deacylation of AA from membrane phospholipids. This is likely mediated through activation of phospholipase A_2, although combined activation of phospholipase C and diacylglycerol lipase is a possible but less likely mechanism. The data presented in these initial studies showed that UVB is capable of inducing release of AA from mammalian cells and suggests that this deacylation is mediated through phospholipase A_2 activation.

Further, UVB was shown capable of inducing generation of various cyclooxygenase AA metabolites. In preliminary studies (11) we have also shown that long wave ultraviolet radiation (UVA - 320-400 nm) induced the release of [^3H] AA from prelabelled C^3H 10T 1/2 cells in a dose dependent fashion.

Utilizing this model we have examined the effect of an exogenous photosensitizer on putative lipase activation (12). Benoxaprofen (2-[4-chlorophenyl]-α-methyl-5 benzoazoleacetic acid) is a non-steroidal antiinflammatory agent which was reported to produce an unusual, immediate type of cutaneous photosensitivity when taken systemically before sun exposure. In experiments designed to examine the mechanism of this photosensitization C3H 10T1/2 cells were prelabelled with [^3H] AA and exposed to benoxaprofen and UVB. The combination of drug and light resulted in release of label into the media. The doses of both drug (5 ug/ml) and UVB (100 J/m^2) sufficient to induce 3 fold stimulation of label release induced no release above control values when either were used alone to treat the cells. These studies suggest that UV-induced alteration of AA metabolism is important not only in the natural sunburn response, but may also play a role in certain types of exogenous photosensitivity reactions.

Mast Cells.

Mast cells are important components of human skin. These cells are capable of releasing the preformed inflammatory mediator, histamine, as well as generating eicosanoids when stimulated. There is evidence that such cells play a role in the inflammatory response of normal and photosensitized skin to UV (13,14). Lim et al. (15) have examined the effect of UV in the presence of the photosensitizer protoporphyrin on eicosanoid metabolism in this cell type. Rat serosal mast cells were pre-labelled with [^{14} C] AA and then treated with protoporphyrin and light in the 400-410 nm range. This resulted in stimulation of release of label as compared to the appropriate controls. TLC examination of media extracts revealed stimulation of production of PGD_2, E_2, $F_{2\alpha}$ and 12-hydroxyeicosatetraenoic acid (HETE).

No deacylation was noted with radiation and uroporphyrin. These findings suggest that eicosanoids derived from mast cells play a role in the photosensitivity associated with erythropoietic protoporphyria (a disease of protoporphyrin excess) but not in porphyria cutanea tarda (a disease of uroporphyrin excess.) Clinically the former is associated with an immediate type of photosensitivity whereas the individuals with the latter condition do not usually develop clinical inflammation on sun exposure.

Macrophages.

Macrophages in vivo synthesize leukotrienes when mammalian systems are under trauma conditions. These eicosanoids act to mediate in part the response of these organisms to trauma. Hardcastle and Minoui (16) examined the effect of UV on macrophages derived from mouse peritoneal exudates. The cells were prelabelled with [^{14}C] AA and treated with either UVC (254 nm) radiation or calcium ionophore. Media extracts were examined by HPLC for identification of leukotrienes. The irradiated cells produced twice as much LTC_4 and four times as much

LTB_4 as the calcium ionophore treated cells. This production was associated with cellular aggregation in the UVC treated cells suggesting that the cells produced these lipoxygenase metabolites when under stress conditions of UV irradiation.

Keratinocytes.

The cell types which intereact most commonly with ultraviolet radiation are, of course, the cells of the epidermis -- keratinocytes, melanocytes and Langerhans cells. The effect of such radiation on eicosanoid metabolism has been examined in keratinocytes derived from human, guinea pig and mouse skin.

We utilized cultured keratinocytes from normal human skin with an experimental protocol as outlined for C3H10T 1/2 cells. The cells were grown until 90% confluent and then prelabelled with [^{3}H]AA (17). UVB induced release of label into media beginning within minutes, with maximum response occurring within the three hours immediately post irradiation. Significant release occurred to UVB doses in the 600-1800 J/m^2 range (see Fig. 1).

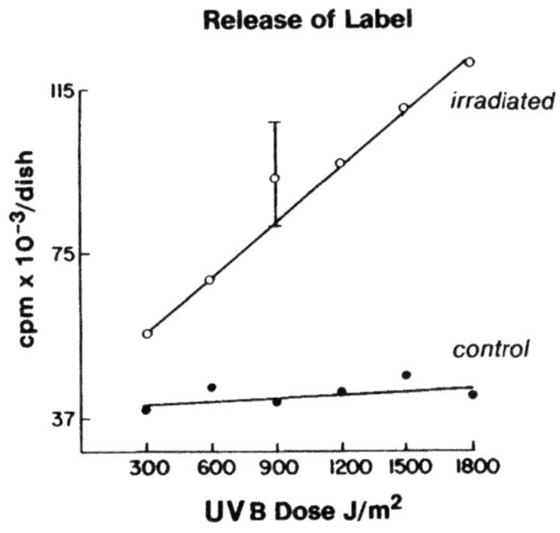

Figure 1. Human keratinocytes were prelabelled with [^{3}H]arachidonic acid and irradiated with UVB. Media were removed three hours post-irradiation and assayed for release of label by scintillation spectorscopy.

The media from control and irradiated cells were extracted and examined for identification of cyclooxygenase products by HPLC with known standards. UVB stimulated the production of both PGE_2 and PGF_2 by the keratinocytes.

In preliminary studies we have also found that UVA induced deacylation of keratinocytes prelabelled with [^3H] AA. We have not, as yet, examined the effect of UVA on AA metabolite production.

Punnonen, Puustinen and Jansen (18,19) also examined the effect of UVB on human keratinocyte eicosanoid metabolism. In their studies they found UVB induced release of AA from prelabelled cells. In contrast to our findings, however, the response was delayed occurring between 2 and 24 hours post irradiation. These authors further assayed the effect of radiation on the labelled cellular lipids of the cultured keratinocytes and discovered that the AA released by UV was predominantly derived from phosphatidylethanolamine. Furthermore a significant quantity of the AA released by UVB treatment was reincorporated into cellular triglycerides. Pretreatment of cells with prednisolone failed to cause more than slight inhibition of UVB mediated release.

In a second communication these authors extended their studies utilizing the same experimental procedure and found UVB stimulated production of PGE_2 as determined by RIA of media extracts at 24 hours after biologically relevant doses of irradiation. In these studies they once again demonstrated UVB stimulated reincorporation into non-phosphorous containing cellular lipids (triglycerides and cholesterol esters). Finally they examined the effect of etretin on the UVB induced membrane change and found no significant effects of this Vitamin A analogue.

Karmali and Safai (20) examined the effect of UV on epidermal cells obtained from guinea pig skin. In these studies cell suspensions from the skin were prepared and the suspensions were fractionated by gradient centrifugation. This procedure separated the cells into three groups - proliferative basal cells, intermediate squamous cells and differentiated squamous cells. These three groups of cells were plated in dishes (short term culture) and then irradiated with a UVB source. Arachidonate metabolite production was assayed by RIA of media extracts. The authors found that UVB stimulated production of a number of different cycloxygenase products (PGE_1, PGE_2, PGF_2, and 6-keto-$PGF_{1\alpha}$) in all three groups of cells. The effect of UV was most marked on the more differentiated cells. By contrast, UV irradiation inhibited thromboxane production by these cells. Time course studies revealed maximal UV stimulation between 6 and 27 hours post irradiation.

Moller et al (21) examined the effect of UVB on leukotriene production by fresh suspensions of mouse epidermal cells derived from BALB/C mice. The authors utilized a fluroscent source and delivered doses of 100-1250 J/m^2. LT production was determined by bioassay for chemotactic activity, reverse phase HPLC identification and RIA. UV failed to stimulate LT production when assayed at times of 10 minutes to 18 hours post irradiation.

SUMMARY

The mechanisms by which ultraviolet radiation induces alteration of various functions of human skin are poorly understood. Such radiation is known to have profound effects on the nuclear material of the cells of the cutaneous organ. These effects have been carefully studied utilizing cells in culture as model systems. Only recently have these systems been utilized to study the effect of radiant energy on membrane metabolism of cells. These early studies, summarized briefly in this communication, indicate that ultraviolet radiation is capable of inducing the deacylation of arachidonic acid from the membrane phospholipids of a number of different types of mammalian cells. This deacylation can occur from irradiation of cells with energy from light sources emitting photons across the entire spectrum (UVA, UVB, AND UVC). The arachidonic acid released can then be metabolized via both cyclooxygenase and lipoxygenase pathways to form many different eicosanoids, dependent on cell type. This effect on eicosanoid metabolism is induced with biologically relevant quantities of ultraviolet radiation and can be induced with UV alone or with UV and chemical photosensitizer. Finally, future studies with these systems should yield information concerning the manner in which the cutaneous organ responds to solar radiation in normal individuals as well as those individuals photosensitized by endogenous and exogenous sensitizers.

REFERENCES

1. Cavallo, J. and DeLeo, V.A. Derm. Clinics 4:181-188, 1986.
2. Epstein, J.H., Fukuyama, K. and Fye, K. Photochem. Photobiol. 12:57-65, 1970.
3. Lowe, N.J. and Stoughton, R.B. J. Invest. Dermatol. 68:134-137, 1977.
4. Rundel, R.D. Photochem. Photobiol. 38:569-575, 1983.
5. Staberg, B., Wulf, H.C., Paulsen, T., Klemp, P. and Brodthagen, H. Arch. Dermatol. 119:641, 1983.
6. Wilis, I., Menter, J.M. and Whyte, J.H. J. Invest. Dermatol. 76:404-408, 1981.
7. Weinstein, I.B., Mufson, R.A., Lih-Syng, L., Fisher, P.B., Laskin, J., Horowitz, A.D. and Ivanovic, U. In: Carcinogenesis: Fundamental Mechanisms and Environmental Effects (Ed. P.B. Reidel), W.B. Saunders, New York, 1980, pp. 543-563.
8. Nondlund, J.J. and Ackles, E.E. Biological Molecular and Clinical Aspects of Pigmentation, University of Tokyo Press, Tokyo, 1985, pp. 501-513.
9. Nondlund, J.J., Collins, C.E. and Rheins, L.A. J. Invest. Dermatol. 86:433-437, 1986.
10. DeLeo, V.A., Hanson, D., Weinstein, I.B. and Harber, L.C. Photochem. Photobiol. 41:51-56, 1985.
11. DeLeo, V.A., Miller, S., Chew, S. and Scheide, S. Photochem. Photobiol. 43:92S, 1986.
12. DeLeo, V.A., Hanson, D. and Scheide, S. Tox. Letters 32:215-220, 1986.
13. Gilchrest, B.A., Soter, N.A., Stoff, J.S. and Mihm, M.C. J. Am. Acad. Dermatol. 5:411-422, 1981.
14. Keahey, T.M., Wehsen, G.F., Kaidbey, K.H. and Lauker, K.M. J. Invest. Dermatol. 78:355, 1982.
15. Lim, H.W., Panker, D. and Marcus, A.J. Clin. Res. 34:763A, 1986.
16. Hardcastle, J.E. and Minoui, S. Manuscript in preparation.
17. DeLeo, V.A., Horlick, H., Hanson, D., Eisinger, M. and Harber, L.C. J. Invest. Dermatol. 83:323-326, 1984.
18. Punnonen, K., Puustinen, T. and Jansen, C.T. Arch. Dermatol. 278:441-444, 1986.
19. Punnonen, K., Puustinen, T. and Jansen, C.T. Photochem. Photobiol. 44:623-627, 1986.
20. Karmali, R.A. and Safai, B. Prostagladins Leukotrienes and Med. 15:277-286, 1984.
21. Moller, A., Schwanitz, H.J. and Czarnetzki, B.M. J. Invest. Dermatol. 87:489-493, 1986.

6

EFFECTS OF ULTRAVIOLET RADIATION ON EICOSANOID METABOLISM IN INTACT SKIN

JAMES S. WALDMAN, M.D. AND HENRY W. LIM, M.D.
Dermatology Section, New York Veterans Administration Medical Center, and Department of Dermatology, New York University School of Medicine, New York, NY 10016, USA

The effects of ultraviolet (UV) radiation on skin have been under intense study in the past decade. This is in part related to the recent large body of knowledge which has accumulated about eicosanoids and their important role as mediators of cutaneous inflammation, including UV-induced cutaneous changes. In this chapter, we will review the known effects of UV radiation on cutaneous eicosanoid metabolism.

THE ULTRAVIOLET RADIATION

Ultraviolet light is composed of wavelengths between two hundred to four hundred nanometers (nm), and further subdivided into shortwave - UVC (200-290nm), middle wave - UVB (290-320nm), and long wave - UVA (320-400nm) components. For these three UV spectra, the energy levels are inversely proportional to the wavelengths, as described by Planck's law : $E = hv = hc/\lambda$, where E = energy of the photon, h = Planck's constant, v = frequency, c = velocity of light, and λ = wavelength. On the other hand, the ability of UV to penetrate into tissue is directly dependent on the wavelength; approximately 1% of UVC, 4-35% of UVB, and 50% of UVA is able to penetrate the epidermis and exert its effect on the dermis [1]. Therefore, it is not surprising that the effects on the skin and eicosanoid metabolism vary according to the spectrum of UV radiation involved.

Sunlight is devoid of wavelengths shorter than UVB, due to absorption of UVC by the ozone layer. It does contain, however, significant amounts of UVB and UVA. UVB is highly erythemogenic and

P. Polgar (ed.), EICOSANOIDS AND RADIATION.

is used therapeutically in photomedicine. UVA is responsible for most endogenous and exogenous photosensitivity reactions, immediate pigment darkening, and it is often used for therapeutic purposes. Visible light (400–760nm), i.e., light detectable by the human eye, does not include UV light; because of its lower energy level, it is of relatively little biologic significance. The majority of photochemical reactions require photon energies of 40 to 120 kcal/mol [1]. The energy levels of UV radiation of 81 kcal/mol of photons (UVA), 95 kcal/mol (UVB), and 102-285 kcal/mol (UVC), along with the penetration of UV into the tissue, explain the important effects of UV radiation on the metabolic activities of intact skin.

The effects of UV radiation on human skin are well known to all who have remained outdoors for extended periods. The classic erythema response to UVB occurs in two phases in animal models: immediate and delayed. In human skin, the immediate erythema is not observed following UVB exposure. The delayed erythema response usually occurs after a latent period of 2 to 6 hours, peaks at 10 to 24 hours, and subsides over the next 72 hours. During this period of resolution, other changes such as desquamation and tanning occur. The time course with UVC is shorter than with UVB; it peaks earlier at 8 hours after exposure, and begins to resolve by 24 hours. UVA radiation frequently results in immediate erythema, which is maximal prior to termination of the exposure period. This erythema persists and gradually resolves over a time period slightly longer than UVB. A monochromatic UVA source may result in a biphasic response; the initial immediate erythema resolving prior to the onset of delayed erythema [1,2].

The clinical appearance of the erythema differs depending on the source of radiation. The shorter wavelength UVC produces a faint red to pink erythema following one minimal erythema dose (MED). The MED unit is defined as the exposure dose that will elicit a perceptible and even erythema at a specified wavelength range. In contrast to UVC, exposure to one MED of UVA or UVB gives rise to a darker erythema.

The histologic changes following UV-exposure have been well characterized, with the appearance of "sunburn cells" as the hallmark of the epidermal alterations. These are dyskeratotic cells with highly eosinophilic cytoplasm and absent or pyknotic nuclei, and are predominantly found in the stratum malpighii at 24 to 48 hours following exposure. They are most prominently seen after irradiation with UVB and UVC. The longer wavelength UVA, because of its lower energy and better penetration, tends to produce much less damage within the epidermis although a few sunburn cells can occasionally be seen. Dermal changes on the other hand, are more notable with UVA. These include endothelial swelling, extravasation of erythrocytes, a neutrophilic infiltrate, and nuclear dust [3].

THE EICOSANOIDS

Eicosanoids are biologically active, unstored, oxygenated metabolites of the twenty-carbon essential fatty acid arachidonic acid [4-7]. Although they have been successfully synthesized or purified only recently [5], the existence of these agents has been postulated by investigators dating back to the initial description of: "Substance H", seen in the inflammatory response characterized by Lewis (1926) [8]; the "Dilator Substance" produced by the epidermis (1945) [9]; and the vasopermeability factor "Leucotaxine" (1936) [9,10]. The currently used nomenclature of prostaglandins, leukotrienes and eicosanoids, were first coined by von Euler [11], Murphy et al [12], and by Corey [5,6] respectively.

Eicosanoids can be classified as autocoids, i.e., substances which are evanescent and exert their effects locally in the microenvironment where they are generated [4]. Such effects in the skin include regulation of vascular tone [13], permeability of capillaries and venules [14-16], epidermal proliferation [17], regulatory effects on the various skin appendages i.e., sweat and sebum secretion [6], along with leukocyte chemotaxis [18-19] and incidence of infections and neoplasms [20-22].

The eicosanoids are synthesized in the cells in response to physical or biochemical stimuli. The precursor arachidonic acid (5,8,11,14-eicosatetraenoic acid - eicosa meaning 20 carbons, tetraenoic meaning 4 double bonds) is the most abundant of the polyunsaturated fatty acids, and is one of the natural constitutents of phospholipids in cell membranes. Present only in its esterified form, it must be hydrolyzed prior to utilization for the synthesis of series-2 eicosanoids. Other natural constituents of cell membranes, 8,11,14-eicosatrienoic acid (dihomo-γ- linolenic acid) and 5,8,11,14,17-eicosapentaenoic acid (EPA) are also suitable enzyme substrates by virtue of their unsaturation at carbon atoms 8, 11, and 14, and are precursors for the less important series-1 and series-3 eicosanoids, respectively.

UV radiation is one of the many stimuli resulting in increased release of arachidonic acid. The mechanism, although it has not been completely elucidated, is felt to be phospholipase activation [23]. Once released by this rate limiting step in eicosanoid generation, the free arachiconic acid can remain within the cell, or diffuse out and interact with neighboring cells [4]. Inside the cell, the free arachiconic acid is rapidly metabolized enzymatically by one of two pathwasy: cyclooxygenase or lipoxygenase.

Cyclooxygenase is a microsomal enzyme which catalyzes the generation of cyclized eicosanoids that include the thromboxanes, prostacyclin (PGI_2), and the other prostaglandins. The enzyme is heme-containing and requires molecular oxygen. The initial step involves formation of an endoperoxide via insertion of two molecules of oxygen (one at C-11, the other at C-15) and the formation of a cyclopentane ring [24]. The hydroperoxyendoperoxide PGG_2 so formed is further reduced to the 15-hydroxy derivative PGH_2, due to the peroxidase-like activity associated with the purified cyclooxygenase [24]. This cyclopentane structure with its two aliphatic side chains ("hairpin-like" configuration) is further classified into groups A to H according to the nature of the functional groups on the cyclopentane ring. The numerical subscript corresponds to the number of double bonds in the side chains of the molecule. Inhibitors of the cyclooxygese enzyme are used clinically and referred to collectively as non-steroidal antiinflammatory drugs (NSAID). They consist of the irreversible inhibitor aspirin, which acetylates the enzyme's active site, and the reversible inhibitors.

The lipoxygenases consist of a group of heterogeneous cytoplasmic enzymes which introduce a hydroperoxide group into arachidonic acid at selective positions. The hydroperoxyeicosatetraenoic acid (HPETE) so formed is reduced to the hydroxyeicosatetranenoic acid (HETE). It is now known that the lipoxygenases can hydroxylate various positions on arachidonate, including 5, 8, 9, 11, 12, and 15 positions [25]. The 5-lipoxygenase generates 5-HPETE which can be reduced to 5-HETE or to the unstable epoxide intermediate LTA_4. Hydrolysis of LTA_4 leads to the formation of LTB_4. Conjugation of LTA_4 with glutathione results in the formation of LTC_4; sequential loss of a glutamic acid residue and a glycine residue leads to the production of LTD_4 and LTE_4.

THE SKIN

The skin is composed of two different layers, epidermis and dermis. Underlying the dermis is the subcutaneous tissue. Although anatomically distinct, they function in-vivo in a unified and interdependent fashion.

TABLE. CONSTITUENTS of SKIN

Epidermis	Keratinocytes
	Melanocytes
	Langerhans Cells
	Merkel Cells
Dermis	Endothelial Cells
	Fibroblasts
	Mast Cells
	Polymorphonuclear Cells
	Lymphocytes
	Histiocytes
	Smooth Muscle Cells
	Nerve Endings

The epidermis is an organized stratified epithelium whose major cellular constituent is the keratinocyte (see Table). The keratinocyte functions as a keratin producing factory which gradually increases its keratin content as it moves from the basal layer to the more superficial layers. In this process it loses its nucleus, becomes non-viable and forms the outermost layer, the stratum corneum. Pentland et al [26] has shown that in vitro keratinocytes synthesized PGE_2, PGD_2, $PGF_{2\alpha}$, and small quantities of 6-keto-$F_{1\alpha}$. These cells have also been stimulated to produce LTB_4 in vitro [27]. Single cell suspensions of murine keratinocytes and epidermal dendritic cells have been shown to generate leukotrienes in response to stimulation with the calcium inophore A23187 and arachidonic acid [28]. Other studies on keratomed epidermis have revealed the presence of 5-lipoxygenase [29] and 12-lipoxygenase [30] activities. Kragballe et al , using homogenized keratomed epidermis containing some papillary dermis, confirmed the presence of 12-lipoxygenase activity, and demonstrated in addition the presence of 15-lipoxygenase activity; the 5-lipoxygenase products (LTB_4, LTC_4, and 5-HETE) were not detected [31]. Aside from the keratinocytes, the other possible eicosanoid gererating cells within the epidermis include: melanocytes, Langerhans cells, and Merkel cells; the eicosanoid profiles of these cell types have not been completely elucidated. In vitro studies with human melanoma cells, however, have failed to show arachidonic acid metabolizing capabilities [6].

The dermis consists of a dense fibrous connective tissue stroma which includes a multitude of different cellular elements: fibroblasts, endothelial cells, mast cells, histiocytes, smooth muscle cells, and specialized nerve endings. This composite of cells is further increased in various cutaneous inflammatory states with the addition of peripheral blood cells: platelets, leukocytes, and erythrocytes. The predominant lipoxygenase enzymes involved in eicosanoid metabolism in platelets is the 12-lipoxygenase, while in eosinophils and neutrophils it is the 15-lipoxygenase and 5-lipoxygenase, respectively. The presence of these various types of cells has made it difficult in ascertaining the cellular source of eicosanoids in in vivo studies. Studies using human skin have shown PGE_2 to be the major prostaglandin produced [32]; on the other hand, PGD_2 was the major prostaglandin generated from guinea pig dermis and epidermis [33]. With regard to epidermal lipoxygenase products, Hammarstrom et al. have shown 12-HETE to be the major product [34]. In the dermis, Kragballe et al have reported 15-HETE to be the predominant lipoxygenase product [35].

PHOTOTOXICITY INDUCED BY EXOGENOUS AGENTS

Phototoxic response occurs in tissues when a molecule of chromophore absorbs radiation of a specified wavelength resulting in a biologic response. This is a dose-related response and can be observed in all tissues exposed to an adequate dose of the phototoxic agent and radiation of the appropriate wavelength. The action spectrum for most known phototoxic agents is in the UVA range. 8-methoxypsoralen, a known phototoxic substance, is frequently used in dermatologic photochemotherapy. The combination of psoralen and UVA light is commonly referred to by the acronym PUVA. It is capable of eliciting a delayed inflammatory reaction clinically resembling sunburn. Its time course is more protracted than UVB erythema; the onset occurs at 12 to 24 hours, and reaches a peak at 36 to 72 hours after exposure. Histologically it resembles the combined effects of UVB and UVA exposures with both epidermal and dermal changes.

Greaves et al in 1978 reported that the levels of PGE_2, $PGF_{2\alpha}$, and arachidonic acid in exudates obtained by suction blister technique from uninvolved skin of PUVA - treated psoriatic patients were comparable to normal controls [36,37]. In the same year, Heiligstadt et al examined the biosynthesis of prostaglandins in the involved skin of nine patients with recalcitrant psoriasis before, during, and after treatment with PUVA; no statistical difference in the levels of PGE_1-equivalents was found [38]. These results were consistent with those of Greaves et al [36,37], and suggest that prostaglandins did not play a role in PUVA-induced inflammation. However, contradicting results were noted by Lord et al in 1976 [39]. They utilized an in vitro skin microsomal fraction preparation and showed that prostaglandin biosynthesis from arachidonic acid was enhanced by the presence of UV radiation and 8-methoxypsorlen (8MOP). Radiation of 254 nm was, surprisingly, more effective than 360 nm. It was further noted that 8MOP stimulated the transformation of PGE_2 into $PGF_{2\alpha}$, which was independent of UV radiation. These results were suggestive of a possible direct effect of 8MOP on PGE_2-9-ketoreductase. These same investigators, utilizing the above mentioned in vitro system, also examined the effects of radiation on eicosanoid generation in the presence of photosensitizing antibiotics [40]. They found an enhancement in prostaglandin biosynthesis of 47-88% over control in the presence of tetracycline, demethylchlortetracycline, and 254 nm radiation. No significant stimulation was seen with 360 nm light. Antibiotics alone had no effect, and other non-photosensitizing antibiotics such as clindamycin, chloramphenicol, and sulfacytine also had no effect. In 1983, Ruzicka et al studied the lipoxygenase and cyclooxygenase activities in mouse skin following exposure to PUVA [41]. They observed a five-fold increase in dermal lipoxygenase activity (expressed as total HETE production). The cyclooxygenase activity, as measured by total prostaglandin production, was unchanged. They suggested that the increased lipoxygenase activity seen in PUVA treated groups was secondary to infiltration of the dermis by inflammatory cells. Using mast cell deficient (W/W^v) mice, Ikai et al showed that the PUVA-induced ear swelling response was not suppressed by topical indomethacin [42]. They also showed that PUVA induced a similar degree of ear thickening response in both control and W/W^v mice, suggesting that mast cells did not play a significant role in this response. Using another phototoxic agent, protoporphyrin, Lim et al demonstrated that irradiation of purified mast cells in the presence of protoporphyrin resulted in the generation of PGD_2 [43].

In summary, the PUVA-induced inflammation appears not to be associated with elevated production of prostaglandins. However, lipoxygenase products, probably generated by the invading peripheral blood leukocytes or platelets, may contribute to this inflammation.

UVA-INDUCED INFLAMMATION

The effects of UVA radiation on human skin have only been studied since 1975, due to the fact that it was not until then that a light source with high UVA output and minimal UVB contamination became available [2,44]. Morrison et al in 1977 examined the effects of the cyclooxygenase inhibitor indomethacin on UVA-induced erythema, and observed that the response was not diminished by topical, oral, or intradermnal administration of this agent [45]. In 1976, Lord et al utilizing an in vitro skin microsomal fraction preparation, observed enhanced prostaglandin biosynthesis from arachidonic acid after UV irradiation of 254 and 360 nm, with the former being a more effective wavelength [39]. Using a suction blister technique, Hawk et al in 1983 detected increased concentrations of arachidonic acid, PGD_2, PGE_2, and 6-keto-$PGF_{1\alpha}$ (the stable prostacyclin metabolite) after UVA irradiation [46]. The peak levels occurred at 5 to 9 hours and returned to control levels by 24 hours. This maximal response in prostaglandin levels was earlier than that observed following exposure to UVB or UVC. Increased histamine levels were also detected following UVA irradiation. The authors suggested that prostaglandins may play a role in the early stages of the UVA-induced inflammatory response.

In summary, an increase in prostaglandin levels has been shown to be temporally related to the appearance of UVA-induced erythema. The role of lipoxygenase products, however, has not been adequately studied.

UVB-INDUCED INFLAMMATION

The role of eicosanoids in UVB-inudced inflammation was first suggested by studies reported in 1970 by Greaves et al, who utilized a Hanovia mercury quartz radiation source with predominant effects in the UVB range [47,48]. Using a continuous skin perfusion technique, Sondergaard et al detected the presence of an ethyl acetate extractable fatty acid from 10 of their 21 patients during the 16-48 hour period after UVB irradiaiton. This fatty acid substance caused a biphasic contractile response of the rat uterus, similar to that induced by PGE_1. No significant increase in kinin activity as compared to control was detected [48]. In another study, the same group of investigators reported increased histamine levels in 6 of the 17 patients following UVB irradiation [47]. However, since it was earlier shown by Partington [49] that the cutaneous response to UV radiation could not be abolished by antihistamines or by prior administration of histamine liberators, the authors suggested that the smooth muscle contracting agent (SMCA) described above was a more important mediator in UVB-induced response.

Although topical applications and intradermal injections of PGE_1 were shown to produce a prolonged cutaneous erythema as early as 1968 [50,51], it was not until 1972 that Mathur et al [52] established the presence of PG in normal human skin (utilizing thin layer chromatography and bioassay), and demonstrated the PG levels to be increased several folds in both dermis and epidermis of albino rats 30 minutes after broad-band UV irradiation (240-370 nm). These investigators suggested that the SMCA previously identified (see above) could be PGE released from the skin in response to UV radiation. That such cutaneous synthesis of eicosanoids could take place was shown by the work of Van Dorp [53] and Ziboh [54] at approximately the same time.

The role of prostaglandins as possible mediators of UVB-induced erythemna was further corroborated by studies using inhibitors of their synthesis. This was partially accomplished in 1974 by Snyder et al in both human and guinea pig skin [55]. Intradermal injections of indomethacin and aspirin (ASA), as well as hydrocortisone (a phospholipase inhibitor), produced a decrease in intensity and delay in the onset of erythema following UVB exposure. Other investigators reported a similar inhibition of erythema with systemic administration of various NSAID [56, 57]. Topically applied indomethacin was also shown to diminish the symptoms of sunburn [58]. In 1976, Snyder further demonstrated that topical indomethacin simultaneously decreased the redness and PGE levels in guinea pig skin to near normal following exposure to 5 MED of UVB [59]. These effects persisted for 24 hours. However, the erythema at 48 hours was unaltered by the application of topical indomethacin, and no elevated PGE level was detected at this time. Therefore, the authors concluded that PGE plays a role in UVB-induced erythema only during the first 24 hours.

In 1976, Black et al examined the time course of PG elevation in human skin 4-48 hours following exposure to 3 MED of UVB, using a suction blister technique [60]. They found a 20-fold increase in PGE_2 and $PGF_{2\alpha}$ levels at 24 hours. Slight increase in activity was detectable at 4 hours while no increase was observed at 48 hours. Bradykinin and histamine levels were not elevated. A similar study by Plummer et al reproduced the above results with respect to $PGF_{2\alpha}$, and in addition, was able to show suppression of both the erythema and rise of $PGF_{2\alpha}$ with topical indomethacin [61]. In 1980, the same London group exposed human subjects to 3 MED of UVB. Utilizing the suction blister technique, they detected increased arachidonic acid, PGE_2, $PGF_{2\alpha}$, and PGD_2 concentrations in the blister fluids of these subjects [62]. Levels were maximal at 24 hours and returned to baseline by 48 hours. The elevated PGD_2 had not previously been shown in human inflammed skin. The source of these eicosanoids, whether it was dermal, epidermal, or invading leukocytes, could not be determined. The same group was able to partially suppress the erythema at 24 hours with oral indomethacin associated with a significant decrease in PGE_2 and $PGF_{2\alpha}$ levels [63]. Black et al also showed that in human skin, the NSAID flurbiprofen partially suppressed the UVB-evoked erythema at 24 hours while completely suppressing the elevation of PGE_2 and $PGF_{2\alpha}$ [64]. No elevation of arachidonic acid, however, was detected. These investigators suggested that this may be explained by the redirection of arachidonic acid metabolism to the lipoxygenase pathway [64]. In 1981, Woodward et al used a guinea pig ear model to evaluate the effects of indomethacin and phenylbutazone on UV erythema [65]. Their results were similar to those of earlier reports, i.e., partial suppression of early erythema. The effects of benzyl-3,5-diacetoxybenzoate (BDAB) on UVB-induced inflammation was studied by Barr et al [66]. BDAB has been previously shown to have an anti-inflammatory activity when applied topically in animal models. However, BDAB was unable to suppress the erythema or the increase in PGE_2 or $PGF_{2\alpha}$ levels after exposure to 3 MED of UVB, suggesting that the suppression of erythema requires the concomitant suppression of the cyclooxygenase products. In 1982, Black et al utilized the suction blister technique to measure the 6-keto-$PGF_{1\alpha}$ levels after exposure to 3 MED of UVB in 11 human subjects [67].

They found the levels to be markedly raised at 6 hours, moderately raised at 24 hours, and normal at 48 hours. Since 6-keto-$PGF_{1\alpha}$ had been shown to possess vasodilatory activity in human skin, they postulated that prostacyclin and 6-keto-$PGF_{1\alpha}$ also played a role in the early phase of UVB-erythema. In 1985, Black et al examined for the first time the lipoxygenase products in human skin after UVB irradiation [68]. 12-HETE and PGE_2 levels in suction blister fluids were elevated at 24 hours, but not at 72 hours after UVB exposure. No significant increase in LTB_4 levels were observed. Since the maximal elevation of 12-HETE was at 24 hours, these investigators concluded that it would be unlikely for 12-HETE to mediate the delayed erythema that persisted beyond the 24-hour period. However, 12-HETE may contribute to the perivenular lymphocytic infiltrate present at 24 and 48 hours after UVB irradiation [3]. In 1986, utilizing a guinea pig model, Burral et al demonstrated an increased transformation of ^{14}C-arachidonic acid into lipoxygenase products (5-HETE and 12-HETE) at 24-72 hours following in vivo exposure to 3 MED of UVB, while the conversion to cyclooxygenase products (PGD_2, PGE_2, $PGF_{2\alpha}$) was decreased [69]. They also demonstrated the presence of a cyclooxygenase inhibitor in the cytoplasmic extract prepared from the guinea pig epidermis obtained at 24 hours after irradiation. This suggested that an endogenous inhibitor of the cyclooxygensase pathway probably contributes to the increased lipoxygenase products at 24-72 hours, and possibly to the prolonged UVB-induced erythema.

In summary, the wealth of data suggests that the levels of all prostaglandins are increased in the initial 24-hour period following UVB irradiation, with levels peaking at 18 to 24 hours, and returning to control values by 48 hours. While there have been only a few studies on the role of lipoxygenase products, available evidence indicates that these metabolites may also contribute to the pathogenesis of UVB-induced inflammation.

UVC-INDUCED INFLAMMATION

There have been only a few studies to examine the pathogenesis of UVC-erythema. A difference between UVB and UVC-induced inflammation was pointed out by the study of Eaglstein et al , in which intradermal indomethacin was shown to suppress erythema induced by UVB but not by UVC [70]. Two years later, Greaves et al , administered 6 MED of UVC to human abdominal skin [71]. Analysis of the suction blister exudates from these subjects revealed that the maximal elevation of the PGE_2, $PGF_{2\alpha}$, and arachidonic acid levels occurred at 18 hours, while the maximal erythema occurred at 12 hours. Oral indomethacin decreased levels of PGE_2 and $PGF_{2\alpha}$ to control levels, but only partially reduced the erythema [72]. Taken together, these results suggest that UVC-induced erythema is only partly mediated by the cyclooxygenase products. Using a hairless mouse model, Ruzicka et al reported a 29% reduction in epidermal HETE production following UVC irradiation [41], however, the biologic significance of this finding remains to be clarified.

ACKNOWLEDGEMENT

Supported by Clinical Investigative Award AR 01595 (HWL) from the National Institutes of Health, and by a Veterans Administration Research Grant.

REFERENCES

1. Kochevar, I.E., Pathak, M.A., Parrish, J.A. In: Dermatology in General Medicine (Eds. T.B. Fitzpatrick, A.Z. Eisen, K. Wolff, I.M. Freedberg, K.F. Austen), McGraw-Hill, Inc., N.Y., 1987, pp. 1441-1451.

2. Parrish, J.A., Anderson, R.R., Ying, C.Y., Pathak, M.A. J. Invest. Dermatol. 67:603-608, 1976.

3. Gilchrest, B.A., Soter, N.A., Stoff, J.S., Mihm, M.C. Am. Acad. Dermatol. 5:411-422, 1981.

4. Marcus, A.J. J. Lipid Res. 25:1511-1516, 1984.

5. Corey, E.J. Ann. N.Y. Acad. Sci. 180:24-37, 1971.

6. Goldyne, M.E. Prog. In Dermatol. 20:1-8, 1986.

7. Moncada, S., Vane, J.R. Br. Med. 39:209, 1983.

8. Lewis, T., Zotterman, Y. Heart 13:203-213, 1926.

9. Blum, H.E. Physiol. Rev. 25:483-530, 1945.

10. Menkin, V. J. Exp. Med. 64:485-502, 1936.

11. Euler, U.S. von J. Physiol. (London) 93:129-143, 1938.

12. Murphy, R.C., Hammarstrom, S., Samuelsson, B. Proc. Natl. Acad. Sci. USA. 76:4275-4279, 1979.

13. Pologe, L.G., Cramer, E.B., Pawlowski, N.A., Abraham, E., Cohn, Z.A., Scott, W.A. J. Exp. Med. 160:1043-1053, 1984.

14. Ford-Hutchinson, A.W. Fed. Proc. 44:25-29, 1985.

15. Soter, N.A., Lewis, R.A., Corey, E.J., Austen, K.F. J. Invest. Dermatol. 80:115-119, 1983.

16. Dowd, P.M., Greaves, M.W. Acta Dermatovener. (Stockh). Suppl. 120:18-22, 1985.

17. Halprin, K.M. J. Invest. Dermatol. 66:339-343, 1976.

18. Dowd, P.M., Black, A.K., Woolard, P.M., Camp, R.D., Greaves, M.W. J. Invest. Dermatol. 84:537-541, 1985.

19. Higgs, G.A., Bax, C.M.R., Moncada, S. In: Leuketrienes and Other Lipoxygenase Products (Eds. B. Samuelsson, and R. Paoletti), Raven Press, New York, 1982, pp. 331-339.

20. Blyth, W.A., Hill, T.J., Field, H.J., Harbour, D.A. J. Gen. Virol. 33:547-550, 1976.

21. Greaves, M.W. Lancet: 189, 1978.

22. Smith, E.B., Mason, W.V., Goodwin, J.S. Lancet: 561, 1978.

23. Blackwell, G.J., Flower, R.J. Br. Med. 39:260-264, 1983.

24. Bakhle, Y.S. Br. Med. 39:214-218, 1983.

25. Kozuka, T., Francis, D.M., Greaves, M.W. Ann. Acad. Med. Singapore 12:87-91, 1983.

26. Pentland, A.P., Needleman, P. J. Clin. Invest. 77:246-251, 1986.

27. Braiun, S.D., Camp, R.D., Leigh, I.M., Ford-Hutchinson A.W. J. Invest. Dermatol. 78:328, 1982.

28. Moller, A., Schwanitz, H., Czarnetzki, B.M. J. Invest. Dermatol. 87:489-493, 1986.

29. Ziboh, V.A., Casebolt, T., Marcelo, C.L., Voorhees, J.J. J. Invest. Dermatol. 80:359, 1983.

30. Hammarstrom, S., Lingren, J.A., Marcele, C., Duell, E.A., Anderson, T.F., Voorhees, J.J. J. Invest. Dermatol. 73:180-183, 1979.

31. Kragballe, K. Desjarlais, L. Duell, E.A., Voorhees, J.J. J. Invest. Dermatol. 87:47-52, 1986.

32. Johnson, C.E., Anggard, E. Scan. J. Clin. Lab Invest. 29:289-295, 1973.

33. Ruzieka, T., Printz, M.P. Biochimica. Biophys. Acta. 71:391-397, 1982.

34. Hammarstrom, S., Hamberg, M., Samuelsson, B., Duell, E.A., Stawiski, M., Voorhees, J.J. Proc. Natl. Acad. Sci. USA 72:5130-5134, 1975.

35. Kragballe, K., Duell, E.A., Voorhees, J.J. Arch. Dermatol. 122:877-880, 1986.

36. Greaves, M.E., Hensby, C.N., Plummar, N.A., Warin, A.P. J. Invest. Dermatol. 71:277, 1978.

37. Greaves, M.W., Hensby, C.N., Plummer, N.A., Warin, A.P. Proc. Br. Pharmacol. Soc. 372P, 1978.

38. Heiligstadt, H., Kassis, V., Weissman, K., Sondergaard, J. Acta. Dermatovener. (Stockh.) 58:213-216, 1978.

39. Lord, J.T., Ziboh, V.A., Poitier, J., Legget, G., Penneys, N.S. Br. J. Dermatol. 95:397-406, 1976.

40. Lord, J.T., Ziboh, V.A., Blick, G., Poitier, J., Kursunoglu, I., Penneys, N.S. Br. J. Dermatol. 98:31-38, 1978.

41. Ruzicka, T., Walter, J.F., Printz, M.P. J. Invest. Dermatol. 81:300-303, 1983.

42. Ikai, K. Danno, K. Horio, T., Narumiya, S. J. Invest. Dermatol. 85:82-84, 1985.

43. Lim, H.W., Parker, D., Marcus, A.J. Clin. Res. 34:736A, 1986.

44. Kaidbey, K.H., Kligman, A.M., J. Invest. Dermatol. 72:253-256, 1979.

45. Morrison, W.L., Paul, B.S., Parrish, J.A. J. Invest. Dermatol. 68:120-133, 1977.

46. Hawk, J.L., Black, A.K., Jaenicke, K.F., Barr, R.M., Soter, N.A., Mallet A.I., Gilchrest, B.A., Hensby, C.N., Parrish, J.A., Greaves, M.W. J. Invest. Dermatol. 80:496-499, 1983.

47. Greaves, M.W., Sondergaard, J. J. Invest,. Dermatol. 54:365-367, 1970.

48. Sondergaard, J., Greaves, M.W. J. Path. 101:93-97, 1970.

49. Partington, M.W. Clin. Sci. 13:425, 1954.

50. Sondergaard, J., Pekka, H., Jorgensen, H.P. J. Path. 109:239-243, 1972.

51. Solomon, L.M., Juhlin, L. Kirschenbaum, M.B. J. Invest. Dermatol. 51:280-282, 1968.

52. Mathur, G.P., Ghandi, V.M. J. Invest. Dermatol. 58:291-295, 1972.

53. Van Dorp, D. Ann. N.Y. Acad. Sci. 180:181-199, 1971.

54. Ziboh, V.A. J. Lip. Res. 14:377-384, 1973.

55. Snyder, D.S., Eaglstein, W.H. J. Invest. Dermatol. 62:47-50, 1974.

56. Miller, W.S., Ruderman, F.R., Smith, J.G. Jr. Arch. Dermatol. 95:357-358, 1967.

57. Gruber, C.M., Ridolfo, A.S., Niekander, R. Mikulaschek, W.M. Clin. Pharmacol. Ther. 13:109-113, 1972.

58. Snyder, D.S. J. Invest. Dermatol. 64:322-326, 1975.

59. Snyder, D.S. Prostaglandins 11:631-643, 1976.

60. Black, A.K., Greaves, M.W., Hensby, C.N., Pazzoglia, A., Plummer, N. Br. J. Dermatol. (Suppl.) 95:21-22, 1976.

61. Plummer, N.A., Greaves, M.W., Hensby, C.N., Black, A.K. Postgrad. Med. J. 53:656-657, 1977.

62. Black, A.K., Fincham, N., Greaves, M.W., Hensby, C.N. Br. J. Pharmacol. 10:453-457, 1980.

63. Hensby, C.N., Plummer, N.A., Black, A.K., Fincham, N., Greaves, M.W. In: Adv. Prostaglandin Thromboxane Res. (Eds. B. Samuelsson, P.W. Ramwell, R. Paoletti), Raven Press, New York 1980, pp. 857-861.

64. Black, A.K., Greaves, M.W., Hensby, C.N. Prostaglandins Med. 5:405-413, 1980.

65. Woodward, D.F., Raval, P., Pipkin, M.A., Owen, D.A.A. Agents Actions 11:711-717, 1981.

66. Barr, R.M., Black, A.K., Mallet, A.I., Greaves, M.W. Prostaglandins Leukotrienes Med. 9:35-43, 1982.

67. Black, A.K., Hensby, C.N., Greaves, M.W. Br. J. Clin. Pharmacol. 13:351-354, 1982.

68. Black, A.K., Barr, R.M., Wong, E., Brain, S., Greaves, M.W., Dickinson, R., Shoot, B., Hensby, C.N. Br. J. Clin. Pharmacol. 20:185-190, 1985.

69. Burrall, B., Ziboh, V.A. J. Invest. Dermatol. 86:643-648, 1986.

70. Eaglstein, W.H., Marisco, A.R. J. Invest. Dermatol. 65:238-240, 1975.

71. Greaves, M.W., Hensby, C.N., Plummer, N.A., Warin, A.P. Proc. Br. Pharmacol. Soc. 445-446P, 1977.

72. Black, A.K., Greaves, M.W., Hensby, C.N., Plummer, N.A., Warin, A.P. Br. J. Clin. Pharmacol. 6:261-266, 1978.

7

PROTECTION AGAINST IONIZING RADIATION WITH EICOSANOIDS

LINDA K. STEEL AND GEORGE N. CATRAVAS

Department of Radiation Biochemistry and Chair of Science, Armed Forces Radiobiology Research Institute, Bethesda, MD 20814-5145.

INTRODUCTION

Prostaglandins (PGs) are extremely diverse in their pharmacological activities. They exhibit both antagonistic as well as cytoprotective properties in the pathogenesis of inflammation. Participation of PGs as chemical mediators in the regulation of immune responses and inflammation are increasingly apparent (1-3). The antagonistic properties of PGs have been implicated in a variety of symptoms resulting from exposure to ionizing radiation. Post irradiation increases in small bowel motility (4), diarrhea, flatulence, abdominal pain (5), mucositis (6), and esophagitis (7,8) have been attributed, in part, to excessive PG production.

In contrast, exogenous PGs, particularly of the E type, have been shown to be cytoprotective against a variety of damaging agents, and a deficiency of endogenous PG has been suggested to contribute to increase susceptibility to injury (9-17). These findings have provided much of the impetus to examine the potential cytoprotective effects of PGs in radiation injury.

Cells in Culture

Perhaps the earliest report of a cytoprotective effect of prostaglandins against ionizing radiation was demonstrated by Prasad (18). PGE_1 (10 μg/ml) added to Chinese hamster ovary (CHO) cells in cultures 1 or 4 hours before x-irradiation increased cell survival and colony forming ability approximately 2-fold, but was ineffective if added 1 or 4 hours after irradiation. He proposed that the observed radioresistance with PGE_1 pretreatment may be through stimulation of adenylate cyclase and subsequent elevation of cylic AMP. Several years later, Lehnert (19) investigated the possible relationship between agents known to elevate cyclic AMP and cell radiosensitivity. Addition of PGE_1 to the culture medium modified the survival of Chinese hamster V-79 cells after x-irradiation by increasing the shoulder portion of the survival curve. Since the compound was not present in the medium at the time of exposure, the possibility of a direct radioprotective effect was ruled out. It was concluded that the effects of PGE_1 are the result of elevated intracellular cyclic AMP, and that cell survival is, in part, dependent on the relationship between these levels and radiosensitivity.

P. Polgar (ed.), EICOSANOIDS AND RADIATION.

Not all reports show PG induced radioprotection. Millar and Jinks (20) found that Chinese hamster cells (line V79-753B) exposed to 1 µg/ml PGE_1 or 10 µg/ml PGA_2 for 20 hours then subjected to gamma irradiation from a ^{60}Co source were not altered in radiosensitivity. The authors did not observe any increase in radiation sensitivity with inhibition of PG synthesis by pretreatment of this cell line with nonsteroidal antiinflammatory drugs (21,22), but glucocorticoid pretreatment increased radiation resistance (22). Hanson and Ainsworth (23) reported that 16,16-dimethyl PGE_2 (2 µg/ml) added to CHO cells 2 hours prior to ^{60}Co gamma irradiation did not confer radioprotection. Similar results were obtained by Holahan and coworkers (24) using exponentially growing synchronous or asynchronous V79-A03 cells pretreated with 14 µM PGE_2 for 2 hours prior to x-irradiation, even though intracellular levels of cyclic AMP were elevated by PGE_2 pretreatment (E. V. Holahan, personal communication). However, PGE_2 did radioprotect asynchronous cells or cells irradiated in the G1 or S phase and exposed to N-ethylmaleimide, a nonspecific thiol-binding radiosensitizer. Recovery was not inhibited by incubation with buthionine-S,R-sulfoximime, a glutathione synthetase inhibitor. These results suggest that glutathione content is not associated with the radioprotective effect of PGE_2. (E. V. Holahan, personal communication). The studies of Millar and Jinks (21,22,25) also suggest a lack of a direct role for elevated glutathione content in the enhancement of radiation resistance of steroid, flurbiprofen or PG-treated cells.

Hinz and Dertinger (26) demonstrated that V-79 cells grown as spheroids could be protected from photon irradiation with PGE_1. However, if these cells were grown as monolayers, no PG-induced protection was conferred. Hanson and coworkers (27) also demonstrated PG-induced radioprotection with 16,16-dimethyl PGE_2 pretreatment of human melanoma (M-1) cells grown as spheroids and exposed to ^{137}Cs gamma radiation. No PG-induced radioprotection was seen in melanoma grown as single cells. These findings suggest a role for cell contact in the protection from radiation treatment and the possibility that membranes of cells in contact may be altered by PG treatment.

Walden and coworkers (28) recently demonstrated a decrease in cellular radiosensitivity of Chinese hamster lung fibroblast (V79A03) cells treated with leukotriene C_4 (LTC_4, 2.5 µM) for two hours prior to 10 Gy x-irradiation. Similar modification of radiosensitivity was demonstrated for LTE_4 but not for LTD_4. This is the first report of a derivative of the lipoxygenase pathway of arachidonic acid metabolism modifying the radiosensitivity of mammalian cells.

Local Irradiation

Clinical studies on the role of PG in radiotherapy are reviewed in another section of this volume. As with cells in culture, there is only tentative or scant evidence in the literature suggesting that PG treatment can protect against local radiation injury. Northway and colleagues (7) evaluated PGE_2 and indomethacin as potential cytoprotective agents of the esophageal mucosa against radiation injury. Opossums received subcutaneous (sc) injection of 5 or 10 μg/kg 16,16-dimethyl PGE_2 1 hour before irradiation and every 4 hours for 24 hours postirradiation; or indomethacin (sc) before or daily for one week postexposure. Animals received 2.25 Gy gamma radiation from a ^{60}Co source, delivered to the entire esophagus in a single dose. PG-treated animals had an increase in the severity of radiation esophagitis, while indomethacin treatment ameliorated the symptoms and severity. Similiar results were seen with aspirin or hydrocortisone treatment (8). They suggested that PG may mediate the inflammatory response resulting from radiation injury and that steroidal or nonsteroidal antiinflammatory drugs, or other pharmacological modulation of endogenous PG synthesis may be of clinical prophylactic value.

In a preliminary report by Delaney (29) intraperitoneal administration of either 16,16-dimethyl PGE_2 or PGE_2 (100 or 250 μg/kg, respectively) to rats one hour prior to receiving 15 Gy abdominal x-irradiation did not alter the mortality curve. Uribe et al. (30) studied oral 16,16-dimethyl PGE_2 administration as a potential protective agent against the damaging effects of local x-irradiation to the abdomen. The mucosa from the small intestine was examined in rats administered PG at 24, 8 and 0.5 hours prior to exposure and 16 and 24 hours following abdominal exposure to 10 Gy. The mucosal crypts and villi, 3 days postirradiation, showed better preservation in rats treated with low doses of 16, 16-dimethyl PGE_2 (100-200 μg/kg). Mucosal damage was exacerbated in animals which received PG doses exceeding 400 μg/kg, similar to the increased severity of radiation esophagitis occurring in PG-treated opossums (7).

The effect of PGE_2 on radiation injury to the small intestine was also examined by Thomas-de la Vega and coworkers (31). Rats were administered 50 μg PGE_2 sc one hour before and 24 hours after 10 or 15 Gy gamma irradiation (^{137}Cs) of an exteriorized segment of the ileum. Morphometric and mitotic (recovery) measurements 3 and 5 days postirradiation demonstrated a cytoprotective effect of PGE_2 on exteriorized, irradiated ileal tissue. The authors suggested that the exogenous addition of PG may correct a postirradiation depletion. This mechanism appears unlikely in view of the reported increases in PG activity in the small intestine following radiation exposure (4). A possible role of PG in intestinal cellular regeneration following radiation exposure has also been postulated (4).

Whole Body Irradiation

Investigators have only recently examined the use of PG as a potential radioprotectant against the effects of whole body exposure to ionizing radiation. In an earlier study, however, Gidali and Feher (32) examined the post-radiation effect of PGE_2 on the regeneration of hematopoietic cells (CFU-S). Gamma-irradiated mice were treated with 0.1 μ g/g PGE_2 24 hours after whole body exposure to 350 rads. The kinetics of CFU regeneration were not improved, even with repeated PGE_2 administration postexposure. These findings suggested that exogenous PGE_2, after radiation insult, does not induce the proliferation of hematopoietic stem cells.

PG administration has been demonstrated to improve intestinal stem cell survival and renewal (33-35), bone marrow CFU-S survival (34,36) and animal survival, if given shortly before radiation exposure. Hanson and Thomas (33) examined the cytoprotective effect of PG on intestinal clonogenic cell survival. Mice were given 16,16-dimethyl PGE_2 sc to the dorsal neck region 1 hour before whole body dose of ^{137}Cs gamma rays. Four days after irradiation, a segment of the jejunum was evaluated for regenerative foci of epithelium. The intestinal clonogenic cells showed an increase in both the shoulder and the slope of the radiation survival curve. The radioprotective effect increased with PG doses ranging from 1 to 10 μ g and plateaued at doses of 10-100 μg. Radioprotection was conferred within 5 minutes of PG administration; maximal protection was observed with administration 1 hour prior to irradiation. No radioprotection of intestinal microcolonies was seen if 16,16-dimethyl E_2 was given 4 or more hours prior to exposure, or given after irradiation. Natural PGE_2 also increased cell survival, but to a lesser degree. No radioprotection was seen with PGE_1 (1-100 μ g) given before exposure. However misoprostol, a synthetic analog of PGE_1, exceeded the radioprotective activity of 16,16-dimethyl PGE_2 (35). Intestinal stem cell survival was increased nearly 600% with 25 μg misoprostol given two hours prior to exposure.

Hanson and Ainsworth (34) also demonstrated radioprotection of bone marrow stem cells (CFU-S) with PG pretreatment. Cell suspensions prepared from femurs of mice previously given 5 μg 16,16-dimethyl PGE_2 sc 1 hour prior to whole body ^{60}Co gamma radiation were injected via tail vein into lethally irradiated recipients. The CFU-S survival (estimated by spleen colony counts 8-10 days post-irradiation) showed an increase in both shoulder and slope of the curve from PG-treated animals. At 50% CFU-S survival, dimethyl PGE_2-induced radioprotection gave a dose modification factor of 1.85. This protection was qualitatively similar to that seen for PG-induced radioprotection of intestinal stem cells (34). The 6-day radiation lethality ($LD_{50/6}$) was also increased 24% in PG-treated mice. The authors suggested that the mechanisms of PGE_2-induced radioprotection of the intestinal cell renewal system and bone marrow CFU-S survival may be through the induction or accumulation of elevated intracellular sulfhydryl compounds (33) and increased intracellular cyclic AMP levels (34). Higher in vivo intestinal radioprotection is conferred with misoprostol, compared to PGE_1, PGE_2 and 16,16-dimethyl PGE_2 (35). These results show that side chain modifications increase the radioprotective properties which contribute to enhanced cell survival.

Recently, Walden and colleagues have examined the radioprotective properties of 16,16-dimethyl PGE_2 on mouse survival (36,37). The extent of radioprotection was dependent on the dose and time of administration. Optimum protection was achieved with 40 µg PG given sc 30 minutes prior to whole-body exposure to gamma radiation from a ^{60}Co source. Under these conditions, a dose reduction factor of 1.72 was achieved. Lesser doses of PG, or earlier administration prior to irradiation were less effective. Similarly, postirradiation injection (5-60 minutes) of 10 µg PG was ineffective in increasing survival. Distribution and metabolism of 16,16-dimethyl PGE_2 in plasma, bone marrow and spleen at 5, 30 and 60 minutes postinjection were also examined. During this time plasma levels of PG remained constant, organ concentrations increased, and most of the 16,16-dimethyl PGE_2 remained unmetabolized. Dimethyl PGE_2 was also found to enhance splenic CFU-S survival in a dose-dependent manner. The authors suggest that the radioprotective benefits of 16,16-dimethyl PGE_2 result from physiologic changes (i.e., hypotension and extravasation) modulated by the synthesis or release of intermediary compounds such as erythropoietin, non-protein sulphydryls and cyclic AMP. Another noteworthy observation was the ineffectiveness of 16,16-dimethyl PGE_2 obtained as an oil dissolved in triacetin, compared to that obtained as an oil and subsequently dissolved in ethanol. Dimethyl PGE_2 dissolved in triacetin was substantially less radioprotective in both animal survival studies and CFU-S survival in spleen. The reasons for these differences are not known; however, they should be considered when interpreting the literature on PG-induced radioprotection.

Modulation of Substrate Availability

The exact mechanisms regulating the release of the 20-carbon essential fatty acid, arachidonate, upon cell stimulation (irradiation or otherwise) are not completely known. The essential fatty acids are precursors of the eicosanoids (prostaglandins, leukotrienes, lipoxins and related substances). They also participate in membrane bilayer fluidity and viscosity, thereby exerting effects on membrane proteins and intrinsic enzymes and molecule translocation or channel formation across cell membranes. Modification or depletion of essential fatty acids has been shown to change phospholipid fatty acid composition, inhibit cyclooxygenase activity and alter both cyclooxygenase and lipoxygenase products (reviewed in 38-41). The modulation of substrate availability by dietary fatty acid modification offers another approach to the evaluation of eicosanoid synthesis in the biological response to radiation exposure.

Cheng et al (42,43) examined the effect of a fat-free diet only, or a daily dietary supplement of 10 mg methyl linoleate or 2% cotton seed oil (approx. 50% linoleate), on fatty acid deficient rats subjected to repeated whole-body x-irradiation of 3 Gy. Fatty acid deficient animals were more radiosensitive than those whose diets were supplemented with linoleate. The authors suggested that the protective effect relates to the correction of the nutritional requirements of essential fatty acids by linoleic acid. Essential fatty acid deficiency has been shown to result in decreased PG production (44,45). Although dietary enrichment with linoleic acid may have effects unrelated to arachidonate metabolism, linoleate supplementation has been shown to increase PG production (46,47).

Intraperitoneal injection of 1 ml of olive oil (mixed glycerides, primarily oleic acid) 1 hour before or 30 minutes after lethal whole-body x-irradiation of mice was found to modestly increase animal survival over the $LD_{50/30}$ of controls (48). The authors speculated that the therapeutic effect of ip administered fats on lethally irradiated animals may partially overcome the malnutrition arising from the inability to utilize orally administered fatty acids. Oleic acid dietary supplements do not cause a change in eicosanoid levels (PGE_2, TxB_2 and LTB_4) in the rat (49).

In a preliminary report (29), Delaney suggests that essential fatty acid deficiency reduces radiation-induced mortality. Fatty acid deficient rats exposed to 14 Gy x-rays or 11 Gy from a 4 MeV linear accelerator in a single dose to the abdomen had increased survival compared to animals on the control diet. The author suggests that the enhanced survival is related to chronic PG depletion by a fatty acid deficient diet.

SUMMARY

More than a decade has past since the first reports suggested that exogenous eicosanoids modify cellular radiosensitivity and increase cell survival. Since then, it has become evident that cells, tissues, and whole organisms differ in radiosensitivity and in the capacity of PG to confer radioprotection. The majority of the studies indicate administration of PG postexposure does not modify radiosensitivity or enhance survival, and in fact may exacerbate symptoms of radiation injury. Exogenous PG administered prior to exposure to ionizing radiation has given mixed results in cell cultures or with local irradiation. Similiarly, studies on the effect of dietary fatty acid modification or depletion on animal radiosensitivity are inconclusive. More recently, whole body irradiation experiments have demonstrated significant PG-induced radioprotection of bone marrow, intestinal stem cells and animal survival. These studies have primarily focused on natural and synthetic PGs of the E series. However, newer synthetic analogues, such as misoprostol, with a longer duration of action (and possibly fewer adverse effects), suggest the potential for improving the radioprotective properties of PGs by side chain modifications.

The mechanism(s) of PG-induced radioprotection are postulated to be through biochemical and physiologic cell changes modulated by some intermediary compound(s). This intermediary substance is most frequently suggested to be cyclic AMP, however erythropoietin, superoxide dismutase, non-protein sulfhydryls, glutathione and others have been proposed. Although the mechanism remains unresolved, PG-induced radioprotection may have important implications for patients receiving radiation therapy for cancer. PGs are secreted by a number of tumors (50-54), and may contribute to radioprotection of malignant cells, reducing the therapeutic benefits of radiotherapy.

References

1. Goodwin, J. S. and Webb, D. R. Clin. Immunol. Immunpath. 15:106-122, 1980.
2. Kunkel, S. L., Chensue, S. W., Fantona, J. C., and Ward, P. A. In: The Reticuloendo-thelial System, A Comprehensive Treatise. Vol. VII: Immunology. (Eds. J. A. Bellanti and H. B. Herscowitz), Plenum Press, New York, 1984, p. 289-301.
3. Trang, L. E. Seminars Arth. Rheum. 9:153-190, 1980.
4. Borowska, A., Sierakowski, S., Mackowiak, J., and Wisniewski, K. Experientia 35: 1368-1370, 1979.
5. Mennie, A. T., Dalley, V. M., Dinneen, L. C., and Collier, H. O. J. Lancet 2:942-943, 1975.
6. Tanner, N. S. B., Stamford, I. F., and Bennett, A. Br. J. Cancer 43:767-771, 1981.
7. Northway, M. G., Libshitz, H. I., Osborne, B. M., Feldman, M. S., Mamel, J. J., West, J. H. and Szwarc, I. A. Gastroenterol. 78:883-892, 1980.
8. Northway, M. G., Eastwood, G. L., Libshitz, H. I., Feldman, M. S., Mamel, J. J. and Scwarc, I. A. Digest. Dis. Sci. 27:923-928, 1982.
9. Mann, M. S. Gastroenterol. 68:946-951, 1975.
10. Robert, A., Lancaster, J. P., Davis, J. P., Field, S. O., and Nezamis, J. E., Scand. J. Gastroenterol. 19:69-72, 1984.
11. Robert, A. Prostaglandins 6:523-531, 1974.
12. Robert, A., Nezamis, J. E., Lancaster, C., and Hanehar, A. J. Gastroenterol. 77:433-443, 1979.
13. Goodwin, J. S. Prog. Cancer Res. Ther. 16:393-415, 1981.
14. Papescu, C. M. Prostaglandins, Leukotrienes and Med. 21:97-105, 1986.
15. Karin, S. M. M. and Fung, W. P. Adv. in Prostaglandin Thromboxane Res. 2:109-111, 1976.
16. Kollberg, B., Nordemar, R., Samuelsson, K., and Bergstrom, S. Gastroenterol. 78:470-485, 1980.
17. Robert, A. Gastroenterol. 69: 1045-1047, 1975.
18. Prasad, N. K. Int. J. Radiat Biol. 22 187-189, 1972.
19. Lehnert, S. Radiat. Res. 62, 107-116, 1975.
20. Millar, B. C. and Jinks, S. Int. J. Radiat Biol. 46:367-373, 1984.
21. Millar, B. C. and Jinks, S. Br. J. Radiol. 54: 505-511, 1981.
22. Millar, B. C. and Jinks, S. Br. J. Radiat. 56: 201-206, 1983.
23. Hanson, W. R. and Ainsworth, E. J. 32nd Ann. Mtg. of Radiat. Res. Soc., Orlando, Florida, March 25-29, 1984, Abst. If-11, p.106.
24. Holahan, E. V., Hagan, M. P., D'Alesandro, M. and Blakely, W. F. 33rd Ann. Mtg. of Radiat. Res. Soc., Los Angeles, CA, May 5-9, 1985, Abst. Ib-11, p.117.
25. Millar, B. C. and Jinks, S. Int. J. Radiat Biol. 47: 539-552, 1985.
26. Hinz, G. and Dertinger, H. Radiat. Envir. Biophys. 21 255-264, 1983.
27. Hanson, W., DeLaurentiis, K. and Malkinson, F. 33rd Ann. Mtg. of Radiat. Res. Soc., Los Angeles, CA., May 5-9, 1985, Abst. He-8, p.97.
28. Walden, T. I. Jr., Holahan, E. V. Jr. and Catravas, G. N. Prog. Lipid Res. 25:587-590, 1986.
29. Delaney, J. Prostaglandins and radiation enteritis. Preliminary report of progress as of 5/1/86. Defense Nuclear Agency (DNA) contract # 001-85-C-0001. Armed Forces Radiobiology Research Institute.
30. Uribe, A., Johansson, C., Rubio, C. and Arndt, J. Acta Radiol. Oncol 23: 349-352, 1984.
31. Thomas-de la Vega, J. E., Banner, B. F., Hubbard, M., Boston, D. L., Thomas, C. W., Straus, A. K. and Roseman, D. L. Surg. Gynecol Obstetr. 158: 39-45, 1984.

32. Gidali, J. and Feher, I. Fifth International Congress of Radiation Research. Seattle, Washington, July 13-20, 1974. Abstract E-27-3 p.292.
33. Hanson, W. and Thomas, C. Radiat. Res. 96:393-398, 1983.
34. Hanson, W. and Ainsworth, E. J. Radiat. Res. 103: 196-203, 1985.
35. Hanson, W. R. and DeLaurentiis, K. Sixth International Conference on Prostaglandins and Related Compounds. Satellite Symposium, Prostaglandins in Gastroenterology: Focus on Misoprostol. Florence, Italy, June 4, 1986, (Abstr.).
36. Walden, T. L., Jr., Snyder, S. L., Patchen, M. L., and Steel, L. K. 34th Ann. Mtg. Radiation Res. Society, Las Vegas, Nevada, Apr. 12-17,1986. Abstr. Cl-11, p. 41.
37. Walden, T. Jr., Patchen, M. and Snyder, S. L. personal communication (manuscript submitted).
38. Mathias, M. M. and Dupont, J. Lipids 14:247-253, 1979.
39. Willis, A. L. Nutrition Rev. 39:289-301, 1981.
40. Morita, I., Takahashi, R., Saito, Y. and Murota, S. J. Biol. Chem. 258: 10197-10199, 1983.
41. Croft, K. D., Codde, J. P., Barden, A., Vandongen, R., and Beilin, L. J. Biochem. Biophys. Acta 834: 316-323, 1985.
42. Cheng, A. L. S., Kryder, G. D., Berquist, L. and Deuel, H. J., Jr. Nutrition 48: 161-182, 1952.
43. Deuel, H. J., Jr., Cheng, A. L. S., Kryder, G. D., and Bingeman, M. E. Science 117: 254-255, 1953.
44. Hansen, H. S. Lipids 16: 849-854, 1981.
45. Dunham, E. W., Belasingham, M., Privett, O. S. and Nickell, E. C. J. Lipid Res. 13 458-467, 1978.
46. Nugteren, D. H., Van Evert, W. C., Soating, W. J., and Spuy, J. H. In: Adv. in Prostaglandins and Thromboxane Research (Ed. B. Samuelsson), Vol. 8, 1980, p. 1793-1796.
47. Epstein, M., Lifschitz, M. and Rappaport, K. Clin. Sci. 63: 565-571, 1982.
48. Ashikawa, J. K. and Anderson, O. K. Radiat. Res. 13: 99-107, 1960.
49. Terano, T., Salmon, J. A., Higgs, G. A., and Moncada, S. Biochem. Pharmacol. 35: 779-785, 1986.
50. Seyberth, H. W., Segre, G. V., Morgan, J. L., Sweetman, B. J., Potts, J. T., Jr., and Oates, J. A. N. Engl. J. Med. 293:1278-1283, 1975.
51. Bockman, R. S. Cancer Invest. 1:489-493, 1983.
52. Karmali, R. A. Cancer J. Clin. 33:322-332, 1983.
53. Cummings, K. B. and Robertson, R. P. J. Urol. 118:720-723, 1977.
54. Plescia, O. J. In: Prostaglandins and Cancer, First Internatl. Conf., (Eds., T. J. Powles, R. S. Bockman, K. V. Horr, and P. W. Ramwell), Alan R. Liss, Inc., New York, 1982, p. 619.

8

PROTECTION OF RADIATION DAMAGE BY EICOSANOIDS -CLINICAL STUDIES

B. WEPPELMANN
Department of Radiation Oncology, University of Alabama at Birmingham, Birmingham, 619 S. 19th Street, Birmingham, Alabama, 35203.

Introduction

Eicosanoids have been implicated in the malignant potential of human and animal tumors. A variety of tumors produce a large amount of eicosanoids, which have been shown to influence metastases and growth but also suppression of the host. (1-11). Radiation therapy of cancer can lead to local tumor control and potential cure. Recent investigations have shown an increase in prostaglandin production and release in cell and animal models by radiation. (12-25).

To explore the interactions between eicosanoids and radiation therapy, the recent literature is reviewed.

Use of eicosanoid inhibitors to modify radiation therapy side effects.

Besides surgery, radiation therapy is the main stay of treatment for head and neck cancers. A potentially curative radical course of radiation therapy consists of 6000 - 7000 rads given in 30 - 35 daily fractions over six to seven weeks. The main side effects of radiation therapy is mucositis of the oral cavity, pharynx and larynx. This is due to loss of the epithelial stem cell population and subsequent break down of the epithelial lining. The process is reversible with time and interruption of radiotherapy is often required to let the mucositis subside. Pain, difficulties in food intake and nutrition are the most destressing symptoms. Tanner (26) studied 25 patients under treatment for various head and neck tumors with radiation and concomitant chemotherapy. Samples drawn before, during and after chemo-radiation therapy were examined using a bioassy against PGE_2. Prostaglandin like material (PGLM), degree of

P. Polgar (ed.), EICOSANOIDS AND RADIATION.

mucositis and stage of treatment were correlated. There was good correlation between the amount of PGLM and the degree of mucositis, increasing with the progression of chemotherapy and radiation therapy. The interpretation of results is hampered by the fact that a large number of patients used Aspirin (ASA) or Paracetamol prior to sampling. But even in view of this, the study demonstrates the involvement of eicosanoids in radiation induced mucositis. Pillsbury (27), in a double-blind trial, investigated the use of Indomethacin to alleviate the mucositis during the radiotherapy for head and neck cancers. The effect of treatment was graded clinically and measured by nutritional parameters. Ten patients received Indomethacin 25 mg QID, eight patients served as control. Radiation therapy was given at 150 rads BID, the minimum dose to the oral cavity being 6000 rads. No difference in mean weight loss between beginning and end of treatment was noted. No differences in treatment success were found, with 40% of controls and 30% of the Indomethacin group locally tumor free. Grade III mucositis was significantly less in the Indomethacin group. The salivary gland flow, however, was better in the treatment group at six months followup. No increase in distant metastases was found in the inhibitor treated patients.

Parotid gland. The role of prostaglandins in the mediation of radiation injury to the parotid gland has been studied by Goldberg (28). Twenty-four hours after radiation therapy, the damage to the parotid gland was measured in rats which had been pre-treated either with Saline, Aspirin or Indomethacin. Damage was expressed by percent of necrotic cells in the removed parotid glands and parotid and serum Amylase levels. Necrosis of the acinar cells appeared in a dose dependent manner. A significant decrease of the necrosis was evident after pre-treatment of rats with Indomethacin, 10 mg/kg and 20 mg/kg and Aspirin 20 mg/kg and 80 mg/kg compared to controls. Serum amylase levels were close to normal after pre-treatment with eicosanoid inhibitors. Interestingly, the protective effect was limited to the acinar cells. Lymphocytes were not protected as seen in the tissue specimen. The dosage of Indomethacin used in this study could probably not be achieved in humans due to GI toxicity. However, 20 mg/kg of Aspirin would be equivalent to roughly 1.5 gram and would certainly be achievable without undue toxicity and still confer significant protection. Other studies have shown a protective effect of WR 2721 (29). This may be mediated through a common pathway, as will be discussed later.

Esophagus. At least 30% of all patients receiving radiotherapy to the chest will develop radiation induced esophagitis. This can range from mucositis to severe damage of muscle layers and subsequent stricture. Northway (30), in 1979, introduced an interesting animal model, the opossum, whose esophagus has a muscular structure similar to humans, with striated muscle in the upper, smooth muscle in the lower and mixed fibers in the mid-portion. Evaluation, after single dose exposure of 1500 to 2250 rads, with the use of a barium esophogram, esophagoscopy and histological sectioning showed typical early changes after seven to 10 days and chronic late changes after four to eight weeks after radiotherapy. Motor functional changes were found by means of balloon manometry to last for up to eight months after the exposure. Using this model, Northway (31) reports the use of Indomethacin 4 mg/kg and 16, 16 Dimethyl PGE_2, 10 μg/kg to modify the esophageal response to 2250 rads of single dose radiation. All animals were examined with esophogram, esophagoscopy and autopsy. Indomethacin treated animals have significantly less mucositis expressed by x-ray exams and endoscopy as control animals. Prostaglandin treated animals have worse mucosal reaction and substantial sloughing of the mucosa. In a separate experiment, Northway (32) compared Aspirin, 600 mg per animal (average of 200 mg/kg) and Hydrocortisone 25 mg/kg with Saline placebo in opossums receiving 2250 rads single dose to the esophagus. The results show clearly the significant protection of the esophageal mucosa by Aspirin and to a lesser degree by Hydrocortisone. The results of Northway have been corroborated by Ambrus (33) in a study on Stump-tailed monkeys. The animals received Sodium Meclofenamate (Meclomen) in doses from 5 mg/kg to 20 mg/kg prior and after exposure to 2000 rads to the upper esophagus. A significant decrease of mucositis was seen after three weeks in the treatment group. The same animals also received radiation to the bladder to investigate the effect of Meclomen on radiation induced cystitis. Animals treated with 20 mg/kg of Meclomen showed significantly less cystitis than control animals. The only clinical study on the patients undergoing radiation therapy for cancer of the lung was reported by Nicolopoulus in 1985 (34). Seventeen patients received 50 mg of Indomethacin BID and 16 patients served as controls. All patients were treated with fractionated radiotherapy of 200 rads per day to a total dose of 4000-6000 rads. Evaluation was performed using a score of symptoms, endoscopy after three weeks of radiation at 3000 rads, and histological examination of biopsy specimens obtained during the endoscopy. The patients in the control group had significantly more symptoms of esophagitis after the fourth week of radiation treatment than the patients in the treatment group. The mucositis scores during endoscopy were significantly better in the Indomethacin group. Histologic examination was more favorable in the treatment group, but not significantly on statistical analysis.

Small and large bowel. Radiation therapy is an important treatment modality for pelvic malignancies, especially cancers of the cervix, prostate and colo-rectum. The dose limiting organ for radical pelvic radiotherapy is the small and to a lesser degree large bowel. Since Bennett (35) postulated that the cholera toxin induces diarrhea by stimulating prostaglandins, the radiation induced small bowel syndrome with its symptoms of diarrhea, cramping, abdominal pain, and nausea and vomiting has been linked to a stimulation of the prostaglandin synthesis and release. Borowska (36), in 1979, examined the effect of 600 rad single dose radiation on the GI motility and PGE_2 content, as measured in a bio-assay in mice receiving Indomethacin 2 mg/kg. A peak in GI motility as measured by the percentage of contrast passed through the small bowel by the animals was observed on the fourth day, only in the control animals. The increase in GI motility was paralleled by the increase in the PGE_2-like activity in control animals with a peak on the fourth day after radiation therapy. In Indomethacin treated animals, the PGE_2 like activity decreased steadily to about 50% of the pre-treatment values.

A large number of clinical studies examined the prevention and treatment of the radiation induced GI toxicity with its very destressing symptoms to the patients. Rauch (37) reported in 1972 the use of sulfasalazine (Azulfidine) in 50 patients with radiation induced colitis. The patient received five grams per day by mouth and 500 mg TID as rectal suppositories. Nine out of 10 patients with mild diarrhea, 12 out of 25 with moderate and 12 out of 15 patients with severe diarrhea were controlled. Overall, 41 out of 50 patients were improved. Goldstein (38) confirmed these results with sulfasalazine in patients with chronic radiation enteritis. He treated 4 patients with refractory chronic diarrhea after radiation therapy for pelvic malignancies with 1 g of sulfasalazine TID. All four patients improved with normalization of small and large bowel appearance on x-rays. Mennie introduced the use of aspirin in the treatment of radiation induced diarrhea implicating prostaglandin release as its cause. The first study (39) reported 15 patients under radiation therapy for cancer of the cervix. The patient received 900 milligrams ASA QID when symptoms appeared. Four out of 15 patients were completely cleared, 8 out of 15 improved. This was followed by the second trial in 1975 (40). Fourteen patients received 972 milligrams ASA (Alkaseltzer) QID, 14 patients placebo. Eleven out of 14 ASA treated patients had significant decrease of diarrhea, compared to 3 out of 14 control patients. Colicky abdominal pain in 8 out of 8 ASA patients improved, compared to none in seven control patients. Stimulated by Mennie's studies, Stryker (41) conducted a randomized trial using Ibubrofen concomitantly with radiation therapy for pelvic malignancies. Seventeen patients received Ibuprofen 400 milligrams QID from the beginning to the end of the radiation therapy course and 16 patients served as controls and were treated with standard therapy when symptoms arose. Daily stool frequency and nausea and vomiting were compared in both groups. In 5 patients in the Ibuprofen group and 6 in the control group, the plasma PGE and PGF levels at the onset of therapy and after five weeks were measured by Radioimmunoassay (RIA). No differences in stool frequency were found. Fifty-nine percent of Ibuprofen patients and 53% of controls had significant diarrhea, requiring additional treatment in 35% and 47% respectively. Nausea was not significantly different in either group, however, 4 out of 15 patients in the control group complained of severe vomiting, compared to 0 out of 17 patients in the Ibuprofen group. PGE and PGF determinations in both groups at the beginning and at the end of the radiotherapy did not show any significant changes, except a slightly higher PGF value in the control group at the end of the radiotherapy course. Lifshitz (42) in 1982 reported measurements of PGE, PGF and 13, 14 Dihydro-15-Keto $PGF_{2\alpha}$ (PGFM) in 17 patients undergoing standard pelvic radiotherapy. Blood was drawn at the beginning, at weekly intervals and six weeks following completion of the radiotherapy. Diarrhea was reported by 76% of patients, nausea and vomiting by 35% and abdominal pain by 53%. The stool frequency increased significantly during the course of radiotherapy. The distribution of PGE, PGF and PGFM did not change during the course of the treatment program and did not show any correlation with clinical events. In a followup investigation, Gal (43) reported that radiation decreased the prostaglandin production of human colon in vitro. Colonic tissue samples from 4 patients were exposed to doses of 500, 1000 and 2500 rads at single dose. PGE_2, $PGF_{2\alpha}$, PGFM, PGD_2, 6-keto$PGF_{1\alpha}$, and TXB_2 were measured after superfusion and mincing by RIA. The decrease of all prostanoids was found after radiation, in a dose dependent manner. Gal concludes that radiotherapy decreases the amount of prostaglandin released by human colon in vitro. This could be caused by the inactivation of the cyclo-oxygenase by free radicals created by absorbtion of x-rays in tissue. A decrease of the prostaglandin then would lead to a decrease in cytoprotection. These findings are in contrast to the current literature, which shows a significant increase in bowel motility and prostaglandin production after radiation therapy. Ludgate (44) in 1985 recommended the use of aspirin in the treatment of radiation induced bowel and soft tissue complications. Ten patients developed complications following radiation therapy for pelvic malignancies, three months to nine years out from the originial radiation therapy. Chronic diarrhea and vaginal vault ulceration subsided after treatment with aspirin 325 mg QID.

Lung tissue. Radiation therapy induced pulmonary fibrosis is a late sequella of radiation therapy to the lung and is apparent in virtually all patients receiving potentially curative doses of radiotherapy to the chest. Acute radiation pneumonitis, characterized by release of surfactant factor from type II pneumocytes (45), is amendable by treatment with steroids. Late appearing radiation fibrosis is not reversible with steroid treatments (46). Karg (47) in 1971 reported significant improvement in side effects of radical radiation for various sites in head and neck and lung cancers. Interpretation of this study is difficult, as the reported 30 patients have different sites of cancer and a multitude of different treatment areas and there is no control group. Klein (48) reported a double blind, randomized trial on patients receiving radiotherapy for breast cancer to the chest wall and supraclavicular area postoperatively. In 59 patients receiving Oxyphenbutazone (Tanderil) for the whole treatment program and 57 patients on Placebo, the amount of radiation induced lung fibrosis, as measured by serial chest x-rays, was followed for two years. After six months, 72% of patients in the treatment group and 89% in the control group had detectable fibrosis. This decreased, after 18 months to 36% and 55%, respectively.

Summary. The involvement of eicosanoids in the mediation of radiation induced changes of the mucosa of the head and neck, esophagus, small and large bowel is complex. Whereas acetylsalicylic acid (ASA) is damaging the gastric mucosa by inhibiting prostaglandin synthesis as shown by Vane (49), prostaglandins can cause diarrhea by increase of small bowel activity. 16, 16 Dimethyl PGE_2 heals gastric mucosal lesions (51), but also worsens radiation induced esophagitis (31). Radiation induced mucositis is caused by death of the stem cells of the basal layer, (52) and subsequent gaps in the mucosal integrity. Whether eicosanoids are directly (as mediators) or indirectly (as indicators) involved, eicosanoid inhibitors are beneficial in ameliorating some of the distressing side effects of radiotherapy to the GI tract. As reviewed here, Indomethacin at 100 mg per day should be effective in preventing severe mucositis and esophagitis. The effect of ASA 900 mg QID on radiation induced diarrhea is established (39, 40). Sulfa-salazine in a dosage of 5 g per day is effective in radiation induced diarrhea and late bowel damage (37). The latter is a condensation product of 5-aminosalicylic acid and Sulfapyridine. It is hydrolized into its components after absorption. The Sulfapyridine component has some bacteriostatic ability thus inhibiting bacterial overgrowth. Presently, this drug is used for the treatment of ulcerative colitis and Crohn disease. (53, 54). The improvement in salivary gland flow reported by Pillsbury (27) and decrease of necrosis of salivary gland cells after protection with Indomethacin reported by Goldberg (28) would be a substantial gain for the patient undergoing radiotherapy for head and neck cancers, as xerostomia is one of the most significant long-term side effects of radiation therapy for head and neck cancers (55). The radio-protector WR 2721 also has been found to protect the salivary gland function (29). A recent report by Donlon (36) suggests that WR 2721 decreases eicosanoid production, thus showing a connection between radiation protection and prostaglandin synthesis and release.

Use of eicosanoid inhibitors to modify tumor response to radiotherapy.

High levels of PGE_2 have been found in head and neck cancers (56). Increased synthesis of PGE_2 and a possible defect in removal of $PGF_{2\alpha}$, especially from larger tumors, have been demonstrated by Karmali (57). In a rat model, Blitzer (58) has shown regression of palate tumors in rats treated with Indomethacin. Four out of 19 rats showed complete remission and 8 out of 19 a partial response. Panje (59) has pioneered the use of Indomethacin as adjuvant in patients with head and neck cancers. Seven patients with advanced and recurrent carcinomas of the head and neck areas received between 75 and 100 mg of Indomethacin per day after biopsy as only treatment modality. The majority of patients had received radiotherapy in the past. At least 2 patients had a complete (100%) response, 3 patients had an 80, 75 and 50% response respectively and 2 patients had stable disease. Macca (60) found that lymphocytes from patients with head and neck tumors showed a decrease in the PHA and CON-A mediated stimulation. This decrease in mitogen stimulation was found to be due to PGE_2 and more pronounced in patients finishing radiotherapy as main treatment modality. In 1/3 of these patients, the mitogen response was restored by the addition of Indomethacin. A similar observation was made by Balch (61) in patients with head and neck cancers. Blomgren (62) reports the inhibition of PHA and PPD response in lymphocyte cultures of patients undergoing radiotherapy for breast cancer. This is due to PGE_2 producing monocytes and will be discussed in a separate chapter in this book by Blomgren.

Direct Antitumor Effects. Eicosanoids have been found elevated in a variety of tumors (63). In patients with breast cancer, prostaglandin production is inversely correlated with the time to recurrence (64, 65). Powles (66), in a phase I-II study, treated patients with breast cancer and acute leukemia with prostaglandin E and prostaglandin A, with no apparent benefits. Ylikorkala reported in two publications (67, 68) increased 6-Keto-$PGF_{1\alpha}$, TXB_2 in patients with ovarian cancer. The use of chemotherapy led to a significant increase in $PGF_{2\alpha}M$. ASA and Indomethacin, given during chemotherapy, did not change the nausea and vomiting induced by chemotherapy, but did decrease TXB_2 generation, thus indicating significant blood levels of the prostaglandin inhibitor used. In several animal studies, the use of prostaglandin inhibitors has resulted in inhibition of tumor growth. Popescu (69) demonstrated the cessation of growth in a rat lymphosarcoma after treatment with 250 mg ASA per kg. In addition, electron microscopy showed disappearance of all malignant characteristics of the examined lymphocytes. Karmali (70) achieved growth retardation in a rat mammary carcinoma model after treatment with 5 mg/kg of either Ibuprofen or Flurbiprofen. The involvement of PGE_2 was demonstrated by observing tumor growth restriction after administration of an anti-PGE_2 antibody. While it is clear that prostaglandins are involved in malignant growth, it is not clear whether the use of anti-prostaglandins will contribute to tumor response and survival of patients treated with chemotherapy or radiation therapy. As discussed by Bennett (63, 64), the use of Benorylate did not reduce the incidence of bone metastases in patients with breast cancer. The results of a double blind study with Flurbiprofen in patients with breast cancer as an adjuvant therapy are not yet available.

Effect of eicosanoid inhibitors and radiation therapy. A beneficial effect on tumor control and survival has been postulated by Ambrus (71). To exclude deleterious effects on tumor control by radiotherapy and concomitant Sodium Meclofenamate to reduce radiation induced esophagitis (33), a transplanted Gardener lymphosarcoma was treated with radiotherapy and 20 mg/kg of Meclomen BID per os. The dose response curves for animals treated with radiotherapy and combination therapy are superimposable. The cure rates after 90 days are superior in the combination group, but not significantly. After cure of the lymphosarcoma with combination therapy, a significantly larger number of tumor cells in the inoculum were required to establish a tumor recurrence. This seems to confer an advantage over conventional radiotherapy.

Do eicosanoids or eicosanoid inhibitors change the sensitivity of cells to radiotherapy? Millar (72, 73, 74, 75) has demonstrated no change in radio responsiveness of V-79 (Chinese hamster) cells. In a recent investigation (72) using a mouse fibrosarcoma, which secretes 2 μg PGE_2 per mg of cells, no change in radiation response after pre-treatment with Flurbiprofen in air or hypoxia could be demonstrated. Northway (32) compared the results of radiotherapy with and without 2.5 mg/kg of Flurbiprofen in mice bearing a transplanted mammary tumor. At high doses of radiation (2500 rads single dose) there was no difference in tumor regression in both groups. At moderate dose levels (1000 rads single dose), the tumor volume in the Flurbiprofen group was significantly smaller, suggesting a combined effect. Macdougall (76) also used Flurbiprofen at 2.5 mg/kg in mice receiving total body irradiation. Using a jejunal stem cell assay and a spleen colony assay, no difference in radiation response of bone marrow stem cells or intestinal crypt cells was found with and without the addition of Flurbiprofen.

Uribe (77) investigated the effect of 16, 16 Dimethyl prostaglandin E_2 and whole body radiation on the small intestinal damage in rats. At doses of 0.1-0.4μg/kg, the villous height and the number of crypts per villous, as expressions of mucosal integrity, were larger than in placebo treated animals. The prostaglandin treated, unirradiated animals, showed a better mucosal integrity than Placebo treated, unirradiated animals, suggesting a direct trophic action of PGE_2. De la Vega (78) performed a similar experiment. Rats received 50μg of PGE_2 one hour before and 24 hours post radiotherapy of 1000 or 1500 rads in a single dose to a surgically exposed ileal segment. Animals treated with PGE_2 had improvement in mucosal thickness, villous height and crypt depth. The number of mitosis in the PGE_2 treated animals was increased, suggesting a higher cell replication of ileal stem cells. Autoradiographically, the uptake of the Iodine 125 labeled Deoxyribose Uridine and triated Thymidine was increased in animals treated with PGE_2 before and after radiotherapy. This indicates a large number of cells in S-phase. As a control group of animals not radiated but treated with prostaglandin E_2 was not used, a direct action, not related to radio protection of PGE, cannot be excluded. One possible explanation is that PGE_2 possibly could advance cells more rapidly into S-phase. As S-phase is a more radio-resistant phase of the cell cycle, this would confer a radio protective action by directly changing the sensitivity of the cell population. Hanson and Thomas (79) and Hanson and Ainsworth (80) also used 16, 16 dimethyl PGE_2 to demonstrate protection of murine intestinal and hemato-poetic stem cells. The D_0 and extrapolation number n (both expressions of direct radio-sensitivity) were increased after pretreatment with prostaglandin E_2. As tumors produce more prostaglandin than normal tissue, endogeneous prostaglandin might indeed protect tumors from the effects of radiotherapy. Weppelmann (81) showed a direct survival advantage for patients with carcinoma of the cervix undergoing potentially curative radiotherapy who received a prostaglandin inhibitor as adjuvant treatment.

In 1963 and 1964, 160 patients with carcinoma of the cervix of all stages underwent potentially curative pelvic radiotherapy. Patients were treated with 4000 rads whole pelvic radiotherapy using a four-field technique with Cobalt 60 radiation. In addition to this, 4000 mg hours of radium were applied into the cervix and uterine cavity. Seventy-six patients were assigned randomly to receive 300 mg of Oxyphenbutazone (Tanderil) daily during the six weeks of radiotherapy and the remainder of patients served as controls. The goal of the study in 1963 and 1964 was to decrease the acute side effects of the pelvic radiotherapy, especially nausea, vomiting and diarrhea and to explore the antiphlogistic properties of this new prostaglandin inhibitors. As seen in Table 1, there were no significant differences concerning the stage distribution, but there were slightly more stage III cases in the control group. The mean age in the patients treated was 53.1 ± 10.6 years and in the controls 54.5 ± 12.6 years.

TABLE 1

Distribution of FIGO Stages

Stage	I	II	III	IV	All
Oxyphenbutazone	13	30	30	2	75
Controls	16	23	43	2	84

Note. Using the chi-square test there were no significant differences.

The survival rates, calculated according to the life table method of Cutler and Ederer, are shown in Figure 1 for both groups. The first four years showed no significant differences between controls and treated patients. The percentages of five-year survival are 70% in the Oxyphenbutazone group and 55% in the controls, significantly ($P<0.05$) different, with the differences still growing. At ten years of survival, we find 62% of the treated patients still living compared to only 44% of the untreated patients, significant at $P<0.01$. The mean survival times 101.8 ± 63 month in the Oxyphenbutazone group and 78.9 ± 58 month in the controls are significantly ($P<0.01$) different.

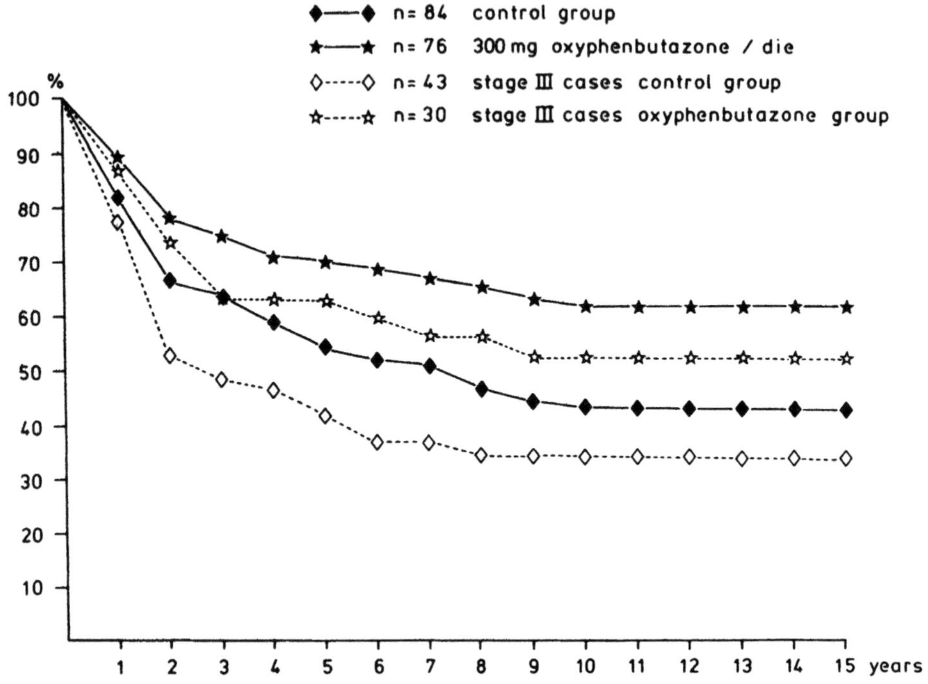

FIG. 1 Survival in Years 1 through 15 after radiation therapy.

The White Blood Cell count and Hemoglobin were unchanged during radiation therapy in both groups. The duration of treatment was nearly equal in both groups. The frequency of cystitis and diarrhea was the same in both patient populations. The medium weight change as a perameter of nutrition was not significantly different. During and after the course of radiotherapy, fistulas between the bladder and vagina and rectum and vagina were seen in the treatment group in 4 cases and in the control group in 2 cases. A long lasting intestinal stricture as sequella of radiotherapy was found in 4 of the 76 patients in the Oxyphenbutazone group and in 7 out of 84 patients in the control group. The differences are not statistically significant. During the first four years of survival, there is a higher percent of survival in the treatment group than in the control group, however, not statistically significant. With increasing time, the survival gain in the treatment arm becomes significant. It is not only true for the overall patient population, but also for patients with stage III which usually have a five-year survival rate of 30-35 percent. Patients with stage III carcinoma of the cervix treated with a prostaglandin inhibitor have a significantly higher survival than patients in the control group. This survival advantage could be explained by either a direct effect of the prostaglandin inhibitor on the resistance of the organism to recurrence, as shown by Ambrus (71) or through restoration of cell mediated immunity as shown by Panje (60) and Blomgren (62).

Summary. The role of eicosanoids in the mediation of the radiation therapy effect still remains a puzzle, although a larger number of pieces now is known. Radiation increases the generation and excretion of prostaglandins (82). Various tumors have a high prostaglandin content (63, 57, 56, 10). There is evidence from animal experiments (77, 78, 79, 80) that eicosanoids can protect from radiation effects on stem cells. This could be mediated by an increase in cyclic AMP, as demonstrated by Prasad (83). Inhibitors of eicosanoid synthesis have been shown to decrease side effects of radiation, to restore cell mediated immunity and to enhance the curative potential of radiation therapy. They also can confer protection from toxicity of hypoxic cell sensitizers, like Mizonidazole (73, 74, 75). These hypoxic cell sensitizers seem to increase the amount of free radicals, mostly O_2^- in the cells (84). Free radicals, on the other hand, have been found to enhance prostaglandin synthesis (85). Radiation protectors, like WR 2721s, are free radical scavengers (86). Recently, Donlon (87) has demonstrated a decrease in radiation induced prostaglandin stimulation by WR 2721. This makes it likely that the effect of radiation therapy and the generation of prostaglandins are mediated through the same mechanism, the free radical (88, 89, 90). Presently, there is enough evidence for the beneficial effect of prostaglandin inhibitors as adjuvant therapy in cancer treatment. At least no evidence has been found to believe that they decrease radiation response (73, 74, 75, 76, 72, 32). Furthermore, there seems to be a direct antitumor effect of prostaglandin inhibitors (91, 58, 59, 92). This should work to the advantage of the patient with cancer. ASA at 325 mg QID (44) or Indomethacin 25 mg TID (59) could be recommended.

References

1. Tisdale, M.J., Expl. Cell. Biol. 51: 250-256, 1983.
2. Smith, B.J., Wills, M.R. and Savory, J., An. Clin. Lab. Sci., 13: 350-365, 1983.
3. Watson, J. and Chuah, S.Y., Eur. J. Cancer Clin. Oncol., 21: 1051-1055, 1985.
4. Elattar, T.M.A., Journal of Oral Pathology, 14: 511-522, 1985.
5. Bockman, R.S., Cancer Investigation, 1, 485-493, 1983.
6. Watson, D.M.A., Kelly, R.W., Hawkins, R.A. and Miller, W.R., Br. J. Cancer, 49: 459-464, 1984.
7. Fantone, J.C., Elgas, L.J., Weinberger, L. and Varani, J., Oncology, 40: 421-426, 1983.
8. Knazek, R.A., Yee, C.L. and Costa, J., Int. J. Cancer, 36: 143-152, 1985.
9. Aitokallio-Tallberg, A., Karkkainen, J., Pantzar, P., Wahlstrom, T. and Ylikorkala, O, Br. J. Cancer, 51: 671-674, 1985.
10. Polterauer, P., Sinzinger, H. and Peskar, B.A., Prostagl. Leukgtr. Med., 22: 249-258, 1986.
11. Keyaki, A., Handa, H., Yamashita, J., Tokuriki, Y., Otsuka, S.I., Yamasaki, T. and Gi, H., J Neurosurg, 61: 912-917, 1984.
12. Eisen, V. and Walker, D.I., Br. J. Pharmacol., 57: 527-532, 1976.
13. Steel, L.K., Sweedler, I.K., and Catravas, G.N., Radiation Res., 94: 156-165, 1983.
14. Hahn, G.L., Menconi, M. and Polgar, P., Prostaglandins and Cancer: First International Conference, pgs. 381-384, Alan Liss, New York, 1982.
15. Walker, D.I. and Eisen, V., Int. J. Radiat. Biol., 36: 399-407, 1979.
16. Trocha, P.J. and Catravas, G.N., Int. J. Radiat. Biol. 38: 503-511,1980.
17. Trocha, P.J. and Catravas, G.N., Advances in Prostaglandin and Thromboxane Research, 7: 851-855, 1980.
18. Ziboh, V.A., Mallia, C. Morhart, E. and Taylor, J.R., Proc. Soc. Exp. Biol. Med., 169: 386-391, 1982.
19. Hahn, G.L., Menconi, M.J., Cahill, M. and Polgar P., Prostaglandins, 25: 783-791, 1983.
20. Steel, L.K. and Catravas, G.N., Int. J. Radiat. Biol. 42: 517-530, 1982.
21. Sinzinger, H., Firbas, W. and Cromwell, M., Prostaglandins, 24: 323-329, 1982.
22. Sinzinger, H. Cromwell, M. and Firbas, W., Radiation Research, 97: 533-536, 1984.
23. Eldor, A., Vlodavsky, I., Hyam, E., Atzmon, R. and Fuks, Z., Prostaglandins, 25: 263-279, 1983.
24. Schneidkraut, M.J., Kot, P.A., Ramwell, P.W. and Rose, J.C., Advances in Prostaglandins Thromboxane, and Leukotriene Research, 12: 107-111, 1983.
25. Schneidkraut, M.J., Kot, P.A., Ramwell, P.W. and Rose, J.C., J. Appl. Physiol., 57: 833-838, 1984.
26. Tanner, N.S.B., Stamford, I.F. and Bennett, A., Br. J. Cancer, 43: 767-771, 1981.
27. Pillsbury, H.C., Webster, W.P. and Rosenman, J., Arch. Otolaryngol., Head Neck, Surg., 112: 552-553, 1986.
28. Goldberg, R.I., J. Am. Dent. Ass., 112: 179-181, 1986.
29. Menard, T.W., Izutsu, K.T., Ensign, W.Y., Keller, P.J., Morton, T. H. and Truelove, E.L., Int. J. Radiation Oncology Biol. Phys.,

10: 1555-1559, 1984.
30. Northway, M.G., Libshitz, H.I., West, J.J., Withers, H.R., Mukhopadhyay, A.K., Osborne, B.M., Szwarc, I.A. and Dodd, G.D., Radiology 131: 731-735, 1979.
31. Northway, M.G., Libshitz, H.I., Osborne, B.M., Feldman, M.S., Mamel, J.J., West, J.H. and Szwarc, I.A., Gastroenterology, 78: 883-892, 1980.
32. Northway, M.G., Bennett, A., Carroll, M.A., Eastwood, G.L., Feldman, M.S., Libshitz, H.I., Mamel, J.J. and Szwarc, I.A., Prostaglandins and Cancer:First International Conference, 799-802, Alan Liss, New York, 1982.
33. Ambrus, J.L., Ambrus, C.M., Lillie, D.B., Johnson, R.J., Gastpar, H. and Kishel, S., Journal of Medicine, 15: 81-91, 1984.
34. Nicolopoulos, N., Mantidis, A., Stathopoulos, E., Papaodysseas, S.,Kouvaris, J. Varveris, H. and Papavasiliou, C., Radiotherapy and Oncology, 3: 23-25, 1985.
35. Bennett, A., Nature, 231: 536, 1971.
36. Borowska, A., Sierakowski, S., Mackowiak, J. and Wisniewski, K., Experientia, 35: 1368-1370, 1979.
37. Rauch, K. and Weiland, H., Strahlentherapie, 143: 660-663, 1972.
38. Goldstein, F., Khoury, J. and Thornton, J.J., Am. J. Gastroenterol, 65: 201-208, 1976.
39. Mennie, ALT., Dalley, V.M., Lancet, 1131, 1973.
40. Mennie, A.T., Dalley, V.M., Dinneen, L.C. and Collier, H.O.J., Lancet, 942-943, 1975.
41. Stryker, J.A., Demers, L.M. and Mortel, R., Int. J. Radiation Oncology Biol. Phys., 5: 2049-2052, 1979.
42. Lipshitz, S., Savage, J.E., Taylor, K.A., Tewfik, H.H. and VanOrden, D.E., Int. J. Radiation Oncology Biol. Phys., 8: 275-277, 1982.
43. Gal, D., Strickland, D.M., Lifshitz, S., Buchsbaum, H.J. and Mitchell, M.D., Int. J. Radiation Oncology Biol. Phys., 10: 653-657, 1984.
44. Ludgate, C.M., J. Assoc. Can. Radiol., 36: 138-140, 1985.
45. Rubin, P., Int. J. Radiation Oncology Biol. Phys., 10: 5-34, 1984.
46. Gross, N.J., Annals of Internal Medicine, 86: 82-92, 1977.
47. Karg, E., Strahlentherapie, 142: 490-493, 1971.
48. Klein, U., Bublath, H. and Heinze, H.G., Strahlentherapie, 144: 421-420, 1972.
49. Vane, J.R., Nature, 231: 232-235, 1971.
50. Hawkey, C.J. and Rampton, D.S., Gastroenterology, 89: 1162-88, 1985.
51. Tarnawski, A., Stachura, J., Ivey, J., Mach, T., Bogdal, J. and Klimczyk, B., Prostaglandins, 21: 147-153, 1981.
52. Hall, E.J., Radiobiology for the Radiologist, 66-78, Harper and Row, Philadelphia, 1978.
53. Svartz, N., Gastroenterolocia, 86: 683, 1956.
54. Vanhees, P.A.M., G.U.T., 22: 404-409, 1981.
55. Mossman, K., Shatzman, A. and Chencharick, J., Int. J. Radiation Oncology Biol. Phys., 8: 991-997, 1982.
56. Jung, T.T.K., Berlinger, N.T., and Juhn, S.K., Laryngoscope 93: 307-312, 1985.
57. Karmali, R.A., Wustrow, T., Thaler, H.T. and Strong, E.W., Cancer Letters, 22: 333-336, 1984.

58. Blitzer, A., Huang, C.C., Arch Otolaryngol, 109: 719-723, 1983.
59. Panje, W.R., Arch Otolaryngol, 107: 658-663, 1981.
60. Maca, R.D. and Panje, W.R., Cancer, 50: 483-489, 1982.
61. Balch, C.M., Dougherty, P.A. and Tilden, A.B., Ann. Surg., 196: 645-650, 1982.
62. Blomgren, H., Wasserman, J. Rotstein, S., Petrini, B. and Baral, E., Radiotherapy and Oncology, 1: 255-261, 1984.
63. Bennett, A., Ann. Clin. Res., 16: 314-317, 1984.
64. Bennett, A., Berstock, D.A., Carroll, M.A., Stamford, I.F. and Wilson, A.J., Advances in Prostaglandin, Thromboxane and Leukotriene Research, 12: 299-302, 1983.
65. Powles, T.J., Advances in Prostaglandin, Thromboxane and Leukotriene Research, 12: 327-331, 1983.
66. Powles, T.J., Coombes, R.C., Depledge, M., Muindi, J. and Powles, R., Prostaglandins and Cancer:First International Conference, 825-830, Alan Liss, New York, 1982.
67. Ylikorkala, O., Kauppila, A. and Viinikka, L., Obstet Gynecol., 58: 483-486, 1981.
68. Ylikorkala, O., Kauppila, A. and Viinikka, L., Gynecologic Oncology, 16: 340-345, 1983.
69. Popescu, M.C., Prostaglandin. Leukotr. Med., 20: 247-254, 1985.
70. Karmali, R.A. and Marsh, J., Prostaglandins Leukotriens and Medicine, 20: 283-286, 1985.
71. Ambrus, J.L., Ambrus, C.M., Halpern, J. and Gastpar, H., Journal of Medicine, 15: 93-103, 1984.
72. Millar, B.C. and Jinks, S., Int. J. radiat. Biol., Vol. 46, No. 4, 367-373, 1984.
73. Millar, B.C., Jinks, S. and Powles, T.J., Br. J. Cancer, 44: 733-740, 1981.
74. Millar, B.C., Chaplin, D.J. and Jinks, S., Advances in Prostaglandin, Thromboxane and Leukotriene Research, 12: 303-308, 1983.
75. Millar, B.C., Jinks, S. and Powles, T.J., Prostaglandins and Cancer: First International Conference, 793-797, Alan Liss, New York, 1982.
76. MacDougall, R.H., Duncan, W. and Clarke, B.M., Prostaglandins and Cancer: First International Conference, 819-823, Alan Liss, New York, 1982.
77. Uribe, A., Johansson, C., Rubio, C. and Arndt, J., Acta Radiologica Oncology, 23: 349-351, 1984.
78. de la Vega, J.E.T., Banner, B.F., Hubbard, M., Boston, D.L., Thomas, C.W., Straus, A.L. and Roseman, D.L., Surgery, Gynecology & Obstetrics, 158: 39-45, 1984.
79. Hanson, W.R. and Thomas, C., Radiation Research, 96: 393-398, 1983.
80. Hanson, W.R. and Ainsworth, E.J., Radiation Research, 103:196-203, 1985.
81. Weppelmann, B., Monkemeier, D., Gynecol. Oncol., 17: 196-199, 1984.
82. Donlon, M., Steel, L., Helgeson, E.A., Shipp, A. and Catravas, G.N., Life Sciences, 32: 2631-2639, 1983.
83. Prasad, K.N., Int. J. Radiat. Biol., 22: 187-189, 1972.
84. Brown, J.M., Cancer 55: 2222-2228, 1985.
85. Taylor, L., Menconi, M.J. and Polgar, P., J. Biol. Chem., 258:

6855-6857, 1983.
86. Fowler, J.F., Int. J. Radiation Oncology Biol. Phys. 11: 665-674, 1985.
87. Donlon, J., Steel, L., Helgeson, E.A., Wolfe, W.W. and Catravas, G.N., Int. J. Radiat. Biol., 47: 205-212, 1985.
88. Gee, C.A., Kittridge, K.J. and Willson, R.L., The British Journal of Radiology, 58: 251-256, 1985.
89. Willson, R.L., Coller, B.S., Lancet, 804, 1984.
90. Dormandy, T.L., Lancet, 1010-1014, 1983.
91. Rubio, C.A., Cancer, 58: 1029-1031, 1986.
92. Young, M.R. and Knies, S., JNCI, 72: 919-922, 1984.

9

RADIATION - INDUCED GENETIC DAMAGE AND THE PROSTAGLANDIN SYSTEM

U. N. DAS AND G. RAMA DEVI

Department of Clinical Pharmacology and Medical Research, Nizam's Institute of Medical Sciences, Punjagutta, Hyderabad-500 482, Andhra Pradesh, I N D I A.

INTRODUCTION

Radiation is one of the sheet anchors of cancer therapy. More often than not radiation is used in combination with anticancer drugs and surgery in the management of cancer. The death of any cell can be observed if the dose of radiation to which it is exposed is sufficiently high. However, after low doses of radiation (precise values vary with the tissue) disturbances in cell proliferation can be observed which may include interference with mitosis and DNA synthesis. This can lead to the formation of cells with abnormal DNA, with qualitative and/ or quantitative abnormalities. These effects of radiation are seen mainly in the tissues undergoing continual renewal such as mucosa, bone marrow, gonads etc., and the spectrum of damage ranging from progressive hypoplesia to total atrophy, depending on the dose. One of the late effects of radiation is the development of cancer.

Mechanism of action

The major biologic effects of radiation occur in three interlinked steps. Initially, photons penetrate the protoplasm, interacting to produce ion pairs. This reaction takes on the order of 10^{-13} s. The second step is a radiochemical reaction of these ions primarily with water, producing free radicals. These reactions take about 10^{-9}s. These free radicals produce a further chain of reactions with themselves and tissue water to produce additional reactive forms such as H_2o_2 and Ho_2. These products, which persist for not more than few seconds, react with critical protoplasmic molecules. This is a low energy reaction system and hence is believed to involve substances of low concentration which are of major importance to the living tissues such as nucleic acids or enzymes.

P. Polgar (ed.), EICOSANOIDS AND RADIATION.

In addition, recent studies revealed that both free radicals and radiation can alter prostaglandin (PG) system as shown in Fig.1

Fig.1: Effect of radiation on essential fatty acid metabolism and PG formation.

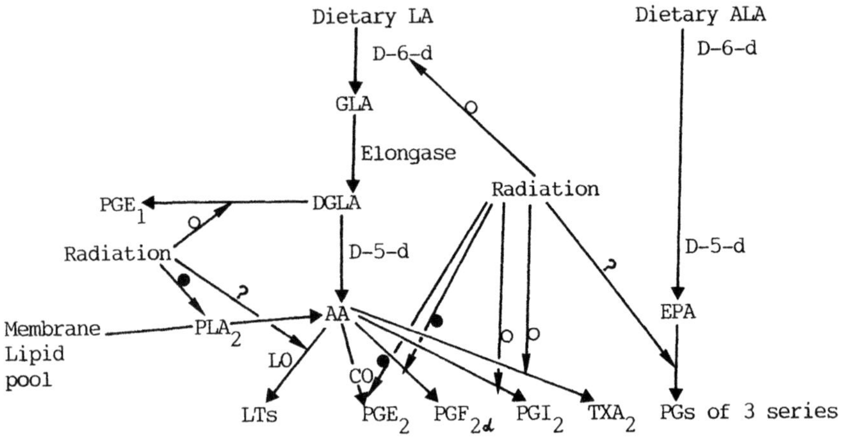

Abbreviations used: PLA_2 = Phospholipase A_2, LO = Lipoxygenase,
CO = Cyclo-oxygenase, D-6-d = Delta-6-desaturase
D-5-d = Delta-5-desaturase, EPA = Eicosapentaenoic acid,
ALA = Alpha-linolenic acid.
o indicates inhibition. ● indicates enhancement.

The well established effects of radiation on PG system include: activation of phospholipase A_2*, increased formation of* PGE_2 *and* $F_{2\alpha}$ *and decrease in the levels of* PGI_2 *and* TXA_2 *(1-3). The inhibitory action of radiation on the enzyme delta-6-desaturase (d-6-d) may lead to a fall in the concentration of GLA in the tissues. Though there are no reports on the effects of radiation on PGs of 1 series, it is likely that, atleast, the level of* PGE_1 *may be low, partly due to inhibitory action on d-6-d and partly as a result of direct action on its formation (for reviews see ref 4,5). Thus radiation can influence both formation of PGs and their precursors in the tissues.*

As both radiation and free radicals are known to damage DNA and alter PG system, we studied the effect of various PGs and their precursors on radiation induced genetic damage (6-9) and a summary of these results are outlined are outlined here.

MATERIALS AND METHODS.

In vitro tests using human lymphocyte cultures

Lymphocyte cultures were done using TC-199 medium with 20% AB serum as described earlier (10). After 48 hours of culture, cell were exposed to 100 r of gamma radiation. One hour after, PGs were added to the cultures as per the treatment schedule. Maintenance and harvesting of the cultures and preparation of the slides for chromosomal analysis were done as per standard procedures and as described earlier (10).

S C E test:

Human lymphocyte cultures were done as described above, except that to obtain visible sister chromatid exchanges (SCEs) bromodeoxyuridine was added, 30μg/5 ml of culture at 24 hour after initiation as described by perry and wolff (11).

In vivo test

Micronucleus test (MN test)

The MN test is a procedure by which one can detect aberrations involving anaphase chromosome behaviour utilising the bone marrow erythroblast. The details of the MN test have been described in one of our earlier papers (6) and was first detailed by schmid (12).

Swiss male mice aged 7-8 weeks and weighing 24-27 gms were used for this study. The animals received 300 r gamma radiation to the whole body. One hour after radiation, PGs were given intraperitoneally in saline. Six animals were used in each group. All the animals were killed 24 hours after the treatment schedule. The bone marrow smears were prepared, stained with May-Grunwald - Giemsa and mounted in Euparol as detailed earlier and as originally described by schmid (12). About 2000 polychromatic erythrocytes and the corresponding normochromatic RBC were scored for the presence of micronuclei for each animal. The ratio of polychromatic to normochromatic (P/N) erythrocytes was used to estimate the effect of the treatments on the bone marrow. A P/N ratio of 1 is normal. Increase in the number of normochromatic RBC is an indication of bone marrow suppression.

Statistics

Chi-square test or studients't test were performed whereever they are appropriate.

RESULTS

At the dose used (300 r to the whole body) gamma radiation was able to induce significant amount of DNA damage (tables 1-3).

The effect of PGE_1, E_2 and $F_2\alpha$ on radiation - induced DNA damage are *given in tables 1-3. It is evident that PGE_1 prevented and/or reversed to a significant extent the genotoxic action of radiation. We have obtained similar results with PGE_1 against BP and DPH (Benzopyrene and diphenylhydantion respectively) induced DNA damage both in vivo and in vitro (6-10, and unpublished data). It can be seen from the data shown in table 1 that gamma radiation induced substantial amount of bone marrow suppression as evidenced by a decrease in the P/N ratio. PGE_1, E_2 and $F_{2\alpha}$ could reduce this suppressive action of radiation on the bone marrow (table 1), suggesting that PGs can help in bone marrow recovery and augment erythropoiesis. But, both PGE_2 and $F_{2\alpha}$ were without any effect on the genotoxic actions of radiation (tables 1-3). In fact, it is evident that both PGE_2 and $F_{2\alpha}$ produced a highly significant increase in the number of chromosomal aberrations and SCEs in lymphocyte cultures in vitro (table 2 and 3). PGE_2 and $F_{2\alpha}$ did not influence the bone marrow cells either alone or in combination with radiation. This may be due to the fact that PGs are normally inactivated in the lungs. Similar to PGE1, thromboxane B_2 (TxB2), metabolite of Tx A2, PGI_2 (prostacyclin) and gamma-linolenic acid (GLA) prevented the genotoxic action of radiation to the bone marrow cells of mice (table 1). GLA can be elongated to form dihomo -gamma-linolenic acid (DGLA), the precursor of PGE1 (Fig.1). Arachidonic acid (AA), the precursor of 2 series PGs, was ineffective against radiation induced damage, but could prevent bone marrow suppression (table 1).*

Discussion

Radiation is known to activate phospholipase A_2leading to the liberation of AA and possibly, that of LA, GLA and DGLA from the cell membrane lipid pool. These fatty acids can be used for the formation of their respective PGs. Radiation is known to enhance the formation of $PGE_2$and F_2 but block the synthesis of TXA_2 and PGI_2 (1-3). It is possible that the effect of radiation on PG synthesis in different tissues may be different. But, in general it seems to increase the production of PGE_2 and $F_{2\alpha}$ since, after radiation there is an increase in the amount of these PGS in the urine. As radiation can block the activity of the enzyme d-6-d, it is possible that over a period of time, once the cell stores of DGLA and GLA are exhausted, there may be a decrease in their formation from LA such that a relative deficiency of 1 series PGs may occur. Since, the cell stores of AA are much larger compared to that of GLA and DGLA the production of 2 series PGs may not be decreased due to decreased availability of the precursor. Thus, over a period of time after exp osure to radiation, the synthesis of PGE_1 may be low with little or no change in PGE_2 and $F_{2\alpha}$production. In addition, radiation can block the synthesis of PGI_2 and TxA_2 from AA. Radiation can also

Table 1: Effect of various PGs and GLA and AA on radiation induced genetic damage to the bone marrow cells of mice.

No.		MN in PRBC	MN in NRBC	P/N ratio
1.	Control	64 (00.53%)	20 (0.17%)	12000/12114 (0.99)
2.	Gamma radiation	1727 (14.40%)***	146 (0.79%)	12000/18452 (0.65)[a]
3.	Radiation +PGE_1	389 (03.24%)***	22 (0.14%)	12000/15280 (0.79)
4.	Radiation +PGE_2	1258 (10.48%)*	64 (0.38%)	12000/16702 (0.72)
5.	Radiation +$PGF_{2\alpha}$	1252 (10.43%)*	66 (0.46%)	12000/14476 (0.83)
6.	Radiation +TXB_2	678 (05.45%)***	62 (0.37%)	12434/16972 (0.73)
7.	Radiation +PGI_2	804 (6.59%)***	270 (2.24%)	12194/12048 (1.01)
8.	Radiation +GLA	692 (05.53%)***	56 (0.38%)	12520/14896 (0.84)
9.	Radiation +AA	1254 (10.03%)*	36 (0.24%)	12504/14730 (0.85)
10.	PGE_1	66 (00.55%)	15 (0.12%)	12000/12184 (0.98)
11.	PGE_2	88 (00.66%)	18 (0.15%)	12000/12110 (0.99)
12.	$PGF_{2\alpha}$	79 (00.66%)	62 (0.51%)	12000/12263 (0.98)
13.	TXB_2	42 (00.34%)	10 (0.08%)	12334/12288 (1.00)
14.	PGI_2	36 (00.29%)	22 (0.18%)	12048/12250 (0.98)
15.	GLA	50 (00.39%)	12 (0.09%)	12678/12704 (0.99)
16.	AA	32 (00.26%)	8 (0.07%)	12216/12174 (1.00)

Dosages: Radiation: 300 r, PGs: $1x10^{-6}M$, AA and GLA: $1x10^{-5}M$

* $P < 0.001$ in comparison to groups 3,6,7,8

** $P < 0.0001$ in comparison to controls.

*** $P < 0.001$ compared to controls.

a indicates bone marrow suppression.

Some of these results were reported in ref. 6.

Table 2: *Effect of prostaglandins on radiation - induced chromosomal aberrations in human lymphocytes in vitro.*

S.No.	Group	Number of aberrations/No. of plates examined
1.	Control	4/100 (1,1,2,0,0)
2.	Radiation	43/100 (10,13, 8, 9, 2)*
3.	Radiation +PGE_1	13/100 (7, 3, 1, 1, 1)**
4.	Radiation +PGE_2	38/100 (9, 11, 8, 6, 4)*
5.	Radiation $PGF_{2\alpha}$	44/100 (21, 6, 6, 6, 5)*
6.	$P\ G\ E_1$	7/100 (2, 0, 3, 2, 0)
7.	$P\ G\ E_2$	23/100 (10, 5, 5, 1, 2)*
8.	$P\ G\ F_{2\alpha}$	25/100 (9, 4, 7, 1, 2)*

Concentrations of PGs used: 0.05 ml of 1×10^{-7} M per culture.

Dose of radiation: 100 r

Mean values of two experiements are given, Values are given as number of aberra/tions/number of plates examined. Types of aberrations are given in the brackets and are in order as: chromatid gaps, isochromatid gaps, chromatid breaks, isochromatid breaks and translocations. No polyploid cells were seen.

* $P < 0.001$ in comparison to control & group 6.

** $P < 0.01$ in comparision to group 2

Table 3: *Effect of prostaglandins on radiation - induced SCE .*

S.No.	Group	SCE per cell
1.	Control	1.52
2.	Radiation	6.40*
3.	Radiation +PGE_1	3.36*
4.	Radiation +PGE_2	7.20*
5.	Radiation +$PGF_{2\alpha}$	7.72*
6.	PGE_1	1.96
7.	PGE_2	6.52*
8.	$PGF_{2\alpha}$	6.69*

Doses of radiation and PGs are same as in table 2. For SCEs, 25 plates were scored and the values are calculated as SCEs per cell. All values are average of atleast 2 experiments.

* $P < 0.01$ in comparison to control and group 6.

** $P < 0.05$ in comparison to radiation.

alter PG metabolism by enhancing lipid peroxidation which is being discussed by Menconi and polgar in this volume.

Free radicals generated due to radiation also can alter PG metabolism, and their effect seem to depend on the amount and type of free radicals available at the site. Deby and Deby - Dupont showed that free radicals can either augument or decrease PG production from AA (14). Hydrogen peroxide at 10^{-4} - 10^{-5} M and hypoxanthine plus xanthine oxidase system at 10^{-4}-5 x 10^{-5}M concentrations seem to augment PG production from AA, where as superoxide dismutase (SOD) at 25 to 50μg/ml inhibit PG synthesis and at higher doses (50μg/ml) facilitates PG formation (14). Free radicals are known to participate in inflammation and excess PGs are found at the site of inflammation which may mean that free radicals, in general, trigger excess formation of PGs.

Free radicals have been shown by Birnboim (15) to have genotoxic actions. Results of our experiments presented here and elsewhere (10) suggest that PGE_2 and $F_{2\alpha}$ can themselves induce damage to DNA, at least, in vitro. From this, it can be suggested that radiation and free radicals bring about their genotoxic actions, atleast partly, by augmenting PGE_2 and $F_{2\alpha}$ levels and by inhibiting the production of beneficial PGs such as TxA_2, PGI_2 and PGE_1. This may also explain why PGE_2 and $PGF_{2\alpha}$ are not effective whereas E1, TxB_2 and PGI_2 could prevent DNA damage due to radiation.

Earlier, it was proposed that a feed back regulation exists between different types of PGs (4, 5, 16, 17): a fall in PGE_1 and TxA_2 synthesis may lead to excess production of PGE_2 and $PGF_{2\alpha}$. Since radiation enhances PGE_2 and $F_{2\alpha}$ synthesis, it is possible that there could occur a fall in the synthesis of PGE_1 and Tx A_2. But, this suggestion is only tentative as we have not measured actual PG levels in the tissues in the present study. It may also be pointed here that there are no studies wherein both 1 series and 2 series PGs were measured simultaneously in tissues exposed to radiation or chemicals.

Precursors of PGs and DNA damage.

It is difficult to use PGs themselves in therapeutics since, they have many actions and are inactivated in the lungs. Hence, we also studied the effect of PG precursors namely AA and GLA, which can be elongated to DGLA, the precursor of PGE1. These results are given table 1 which showed that GLA but not AA can prevent radiation induced genetic damage to the bone marrow cells of mice. Similar results were obtained with AA and GLA when tested against BP -

induced genetic damage (6). In vitro studies have also revealed that AA by itself has genotoxic actions (unpublished data) to human lymphocytes, whereas LA, GLA and DGLA behave similar to PGE1. Studies done by Emerit al (18) showed that PMA-induced chromosomal damage to lymphocytes in vitro can be prevented by inhibitors of AA metabolism, suggesting indirectly a role for 2 series PGs and for PG synthetase dependent events in genetic damage. The results of the present investigation and of those published elsewhere (10) are in accord with this. But, our unpublished results suggest that AA induced chromosomal damage can be blocked by SOD, indicating a role for superoxide radical and possibly other free radicals.

Free radicals and PGs in DNA damage.

In this connection, It is interesting to note that PGE_1, PGE_2 and $F_{2\alpha}$ can modify superoxide release by stimulated neutrophils (19,20). PGE_1, E_2 and D_2 are capable of inhibiting FMLP - activated O_2^- generation in a concentration dependent manner (10^{-6} M - 10^{-9} M): 17% inhibition at 10^{-8} M to 70% inhibition at 10^{-6} M, where as $PGF_{2\alpha}$ was not effective and reduced O_2^- generation by only 18% at 10^{-6} M (19). The inhibitory effect of PGs on O_2^- generation correlated well with increments of cyclic AMP in neutrophils. But none of the PGs reduced PMA or STZ - induced neutrophil activation significantly suggesting that the inhibitory action of PGs depends on the stimulus used. The action of PGE_1, E_2 and $F_{2\alpha}$ on genetic damage induced by radiation is unlikely to be due to their action on free radical generation since, E_1 and E_2 inhibit O_2^- generation where as E_1 and E_2 have opposite actions on genetic damage.

The possibility that PGs are modifying genetic damage by altering cyctic AMP levels and/or by the activities of DNA and RNA polymerases can not be discounted at present. To our knowledge, there are no reports on the influence of cyctic AMP cyclic GMP on genetic damage. Our preliminary results showed that cyclic AMP can prevent DNA damage induced by radiation.

Genetic damage, PGs, oncogenes and Cancer.

Whatever may be the mechanism (s) of action of PGs, the very fact that PGs are able to modify genetic damage itself is interesting. Further, tumour cells are known to contain excess of PGE_2 and $PGF_{2\alpha}$(21, 22) and several mutagens and carcinogens have been reported to stimulate $PGE_2/F_{2\alpha}$synthesis in the cells (23). In all these studies, there is no mention of PGE1 synthesis or levels. If, our earlier hypothesis (4) is true, it is likely, that PGE_1 levels could be low in the tumour cells and its synthesis is reduced by carcinogens and mutagens.

But, this remains to be established. PGs are not stored in the cells and are formed de novo. Thus, it is likely that the excess of $PGE_2/F_{2\alpha}$ formed in the tumour cells or in response to mutagens and carcinogens are released into the intercellular milieu and may act as diffusible clastogenic factors (6).

A strong link is believed to exist between the location of cellular oncogenes and the chromosomal aberrations observed in various forms of cancer. Furthermore, it has also been suggested that fragile sites on chromosomes probably predispose to chromosomal rearrangement and thus, to human neoplasia. The association between oncegenes, chromosome aberrations and heritable fragile sites in cancer have recently been reviewed (24, 25). Since PGs and their procursors can influence chromosomal aberrations and thus chromosomal rearrangement, it is possible that an alteration in the PG system can activate/inactivate oncogene expression and ultimately the development of cancer (26).

Since GLA and DGLA can prevent DNA damage both in vivo and in vitro and AA is genotoxic in nature, it will be interesting to use GLA and DGLA to prevent DNA from the genotoxic actions of radiation therapy. This is supported by our recent study wherein we showed that oral administration of evening primrose oil, a rich source of LA and GLA, can prevent BP - induced genetic damage in mice (27). In another set of investigations, we also showed that GLA, DGLA and AA can selectively kill tumour cells in vitro (28) possibly, by augmenting free radical generation. Even lymphokines such as interferon and tumour necrosis factor also seen to mediate their actions by inducing the release of LA, GLA and AA from the cell membrane lipid pool and by enhancing free radical generation only in the tumour cells (29, 30). From these evidences, it can be suggested that judicious administration of LA, GLA, DGLA, AA and other similar fatty acids and PGs may help to prevent DNA damage and oncogene activation in normal cells due to carcinogens and mutagens and at the same time may selectively kill tumour cells by enhancing free radical generation and lipid peroxidation and by augmenting the actions of lymphokines as shown in fig.2. Further studies are needed to verify these interesting possibilities and are now in progress in our laboratory.

REFERENCES:

1. *Allen, J.B., Sagerman, R.H., Stuart, M.J. Lancet. ii: 1193 - 1196, 1081.*

2. *Setty, B.N.Y., Jurek, E., Ganley, C., Stuart, M.J.Prostaglandins. Leukotrienes. Med. 14 : 205 - 213, 1984.*

3. *Polgar, P., Taylor, L. Prostaglandins, 19: 693-700, 1980.*

4. Das, U.N. Med. Hypotheses. 11: 185-194, 1983.

5. Horrobin, D.F. Med. Hypotheses, 6: 687-709, 1980.

6. Das, U.N., Devi,G.R., Rao., K.P., Rao, M.S. Prostaglandins. 29: 911-920, 1985.

7. Das, U.N., Devi, G.R., Rao, K.P., Rao, M.S. Nutrit. Res. 5: 101-105, 1985.

8. Das, U.N., Devi G.R., Rao, K.P., Rao, M.S. IRCS. Med.Sci.11:863-864, 1983.

9. Das, U.N. IRCS. Med.Sci.11:300-301, 1983.

10. Devi, G.R., Das, U.N., Rao, K.P., Rao, M.S.prostaglandins. Leukotrienes. Med. 15:109-113, 1984.

11. Perry, P.E., Wolff, S.Nature, 251:156-158, 1974.

12. Schmid, W.Mutation. Res.31:9-15, 1975.

13. Bailey, J.M.In; Lipid metabolism in mammals (Ed.F.Snyder), Plenumpress, New York, Vol.2, 1977, P 352.

14. Deby, C., and Deby-Dupont, G.In: Biological and clinical aspects of superoxide and superoxide dismutase (Eds.W.H.Bannister and J.V.Bannister), Proc.Fed.Eur.Biochem.Soc.Symposium No:62, 1982, P 84-97.

15. Birnboim, H.C.Science. 215:1247-1249, 1982.

16. Das, U.N.Int.J.Tiss.Reac. 2:233-236, 1980.

17. Das, U.N.Int.J.Tiss.Reac. 3:89-93, 1981.

18. Emerit, I., Levy, A., Cerutti, P.Mutation.Res.110:327-335, 1983.

19. Sedgwick, J.B., Berube, M.L., Zurier, R.B.Clin.Immunol. Immunopathol. 34: 205-215, 1985.

20. Penfield, A., Dale, M.M.FEBS.Lett.181:335-338, 1985.

21. Bennett, A., Carroll, M.A., Stamford, I.F., et al.Br.J.Pharmacol. 74:207-212, 1981.

22. *Levine, L.Adv.Cancer. Res.35:49-79, 1981.*

23. *Levine, L. Nature. 268:447-448, 1977.*

24. *Rowley, J.D. Nature. 301: 290-291, 1983.*

25. *Rowley, J.D. Nature. 308:607-608, 1984.*

26. *Das, U.N. Speculations. Sci. Tech. 9:99-102, 1986.*

27. *Das, U.N., Devi, G.R., Rao, K.P. IRCS. Med.Sci. 13: 316-317, 1985.*

28. *Begin, M.E., Das, U.N., Ells, G., Horrobin, D.F. Prostaglandins. Leukotrienes. Med.19: 177-186, 1985.*

29. *Das, U.N., Huang, Y.S., Begin, M.E., Horrobin, D.F. IRCS. J.Med.Sci. in press.*

30. *Das, U.N., Ells, G., Begin, M.E., Horrobin, D.F. J.Free. Radicals. Biol. Med. in press.*

Note: For further details about Fig. 2. please see the following references:

Das, U.N., et al. J. Free Radicals Biol. Med., in press.

Das, U.N., et al. IRCS J. Med. Sci., in press (1986).

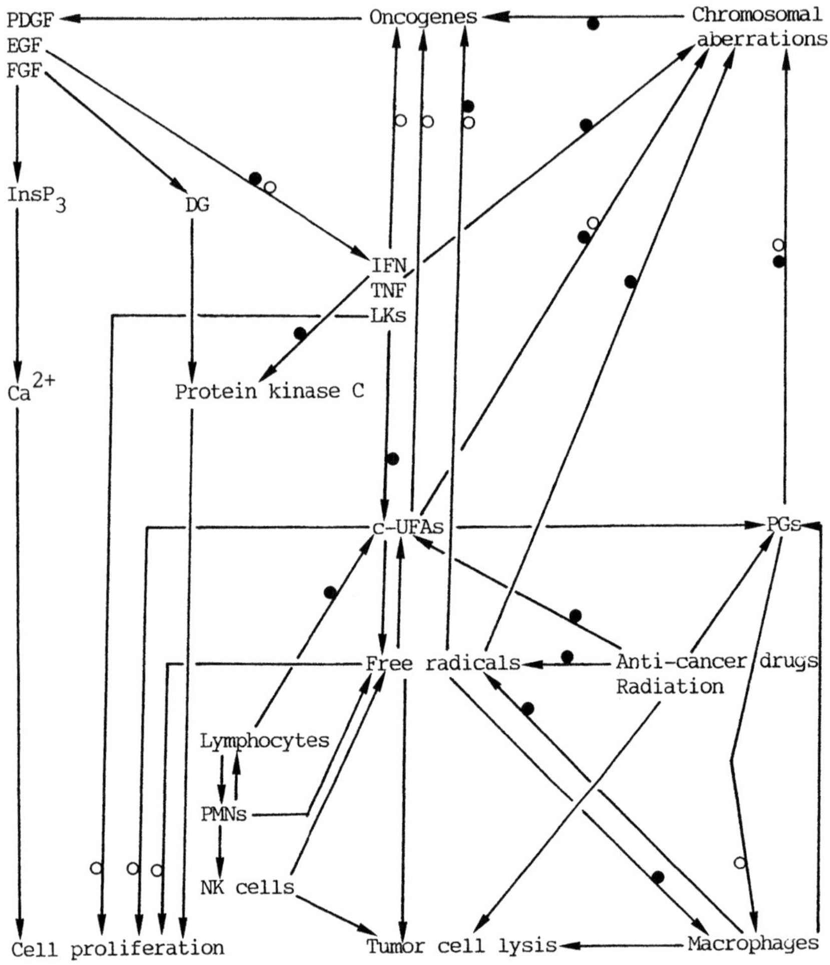

Fig. 2. Scheme showing possible interaction between radiation, anti-cancer drugs, PGs, c-UFAs, free radicals, lymphokines, and tumor cells.

PDGF = Platelet derived growth factor, EGF= Epidermal growth factor, FGF = Fibroblast growth factor, IFN = Interferon, TNF = Tumor necrosis factor, LKs = Lymphokines, DG = Diacylglycerol, $InsP_3$ = Inositoltriphosphate, c-UFAs = cis- Unsaturated fatty acids, PGs = Prostaglandins, PMNs = Polymorphonuclear leukocytes, NK cells = Natural killer cells.

○ indicates negative influence or inhibition of action.

● indicates positive influence such as increased release, augmentation of response or activation.

10

Radiation, Lipid Peroxidation and the Role of Oxygen Radicals in Eicosanoid Metabolism

Michael Menconi and Peter Polgar
Boston University School of Medicine
Department of Biochemistry
Boston, Massachusetts 02118

Introduction

Exposure to ultraviolet or ionizing radiation can initiate the production and/or release of potent metabolites within specific organs, tissues or cell types, resulting in a wide range of effects. One class of biologically active agents whose production is affected by radiation is the eicosanoids. The biosynthesis of eicosanoids from their fatty acid precursors can be stimulated by either specific cell surface membrane receptors or general cell injury. Numerous studies have demonstrated that radiation alters the biosynthesis of prostaglandins (PGs) in a wide variety of tissues, including the lung, spleen, liver, skin and vasculature. While the precise role of PGs in the biological response to radiation is not fully understood, these alterations may be related to subsequent normal as well as pathophysiological responses.

The synthesis of eicosanoids involves free radical fatty acid intermediates and is sensitive to the cellular levels of peroxide molecules. Radiation is known to generate radicals and hydroperoxides. Although the exact mechanism(s) by which radiation alters eicosanoid biosynthesis is not fully understood, the generation of oxygen radicals and hydroperoxides may be an important constituent. This chapter will discuss the role of radicals and hydroperoxides in radiation-induced alterations in eicosanoid biosynthesis.

Radiation-Induced Production of Free Radicals and Lipid Peroxidation

The effect of radiation on tissues is primarily based on the reactions of radicals produced by the radiological disintegration of water with the surrounding biologically important molecules. In the absence of oxygen, irradiation of water produces a wide variety of reactive products, including hydroxyl radicals, hydrogen radicals, hydrated electrons, and hydrogen peroxide (1). In the presence of oxygen, hydrated electrons can react with oxygen to generate the superoxide anion radical (2). The superoxide radical may react with membranes in several ways, including participating in redox reactions with various cations bound to the membrane surface (3) or oxidizing sulfhydryl residues in membrane proteins (4, 5). On the other hand, the chemical dismutation of superoxide radical is reported to result in the formation of singlet oxygen (6, 7, 8), which is a much more efficient initiator of hydroperoxide formation of membrane unsaturated fatty acids than ground state oxygen (9, 10). The enzyme superoxide dismutase catalyzes the dismutation of superoxide radical to ground state oxygen and H_2O_2 (7, 8, 11, 12). This enzyme may therefore be expected to protect membranes from oxidative damage by either superoxide anion radical or singlet oxygen and its protective action against biochemical peroxidation has been demonstrated in a number of membrane systems (13, 14, 15, 16, 17). In the presence of metal ions, superoxide and H_2O_2 can react to produce hydroxyl radicals by a series of reactions known as the Fenton reactions (18). The hydroxyl radical is a powerful oxidizing agent and reacts with a wide variety of organic molecules (2). Because of its high reactivity, the hydroxyl radical can cause biological damage if it is produced close to a critical site. This radical is accepted as playing a leading role in producing radiation damage (19).

The generation of oxygen-derived free radicals and peroxide molecules induced by radiation results in the peroxidation of membrane lipids (20). Polyunsaturated acyl chains of membrane phospholipids are especially sensitive to lipid peroxidation (16, 21). The details of the exact mechanisms by which radiation induces lipid peroxidation have yet to be worked out. However, the chain reaction leading to lipid peroxidation appears to involve the loss of the a-methylene hydrogen atom in the fatty acid hydrocarbon chain to produce a free radical, bond migration to give a conjugated diene, hydroperoxide formation, ring formation, and chain cleavage to produce malondialdehyde (22, 23, 24, 25). This scheme involves predominantly the unsaturated fatty acids and indicates that the greater the degree of unsaturation, the greater the degree of the fatty acid undergoing peroxidation.

The irradiation of phospholipid micelles or liposomes induces the accumulation of hydroperoxides and conjugated diene bonds, and the formation of various end products of lipid peroxidation such as malondialdehyde (26, 27, 28). As typical of chain free-radical reactions, the yield of lipid peroxidation increases with increasing radiation dose and decreasing dose rate (26, 29). Mooibroek et al (27) examined the radiosensitivities of several fatty acids dispersed under different conditions and found that in single component micelles the ease of peroxidation was linoleic (18:2) > linolenic (18:3) > arachidonic (20:4) > docosahexaenoic acid (22:6), whereas in mixed micelle or liposomal dispersions this order of reactivity was reversed. Petkau and Chelack (16) found that x-irradiation of unsaturated phosphatidylcholine dispersions resulted in peroxidation of the lipid and superoxide dismutase partially prevented the peroxidation.

Konings and coworkers demonstrated that liposomes prepared from lipids extracted from rat liver microsomal membranes were susceptible to radiation-induced peroxidation (30). These authors observed a radiation-induced increase in peroxide content accompanied by a decrease in the proportion of polyunsaturated fatty acids in the membrane. Glutathione (10^{-4} M) and cysteamine (10^{-4} M) partially prevented the peroxidation while the lipid-soluble antioxidant a-tocopherol (10^{-5} M) prevented peroxidation completely. Wills and Wilkinson (31) found that microsomes, lysosomes, mitochondria and nuclei extracted from rat liver were peroxidized by irradiation, with the microsomes being the most radiosensitive. In the presence of oxygen, peroxides continued to be formed after irradiation if the irradiated samples were incubated at 37°C. Although radiation-induced peroxidation of unsaturated lipids has been well documented in artificial membrane systems, similar events are probably occurring in vivo.

Structural and Functional Consequences of Lipid Peroxidation

Irradiation of lecithin liposomes brings about structural and permeability alterations which are usually attributed to lipid peroxidation. Fluorescence polarization measurements, upon irradiation, indicate an increase in the fluidity of the head-group region of the lipid bilayer and a decrease in fluidity and increase in polarity within the hydrophobic core of the bilayer (32). Similar effects on membrane fluidity were observed in membrane lipids upon irradiation of cells or membrane preparations. For example, fluorescence polarization measurements on human erythrocyte ghosts administered relatively high doses of gamma-radiation (0.25 to 2 kGy) indicate a decreased fluidity and an increased polarity of the hydrophobic interior of the lipid phase of the membrane (33, 34). Such alterations are consistent with the postulated introduction of polar head groups (hydroperoxides and aldehydes) into the hydrocarbon regions of the bilayer and the excision of double bonds as a result of lipid peroxidation. Nakazawa and coworkers (26, 35) found that the permeability to glucose of irradiated liposomes or of liposomes made of irradiated lipids increases. A parallel course was observed in the dose response of glucose release and malondialdehyde formation (lipid peroxidation) in irradiated soybean lecithin liposomes. Both processes showed a common dose threshold at about 20 Gy and a parallel inhibition by various radical scavengers (35).

Considerable evidence suggests that lipid peroxidation may lead to structural and functional alterations in membrane proteins, particularly membrane-bound enzymes. Reports have indicated that alterations occur in the activities of such membrane-bound enzymes as mitochondrial ATPase (36), Na^+-ATPase (37) and collagenase (38, 39). Several studies have demonstrated that radiation-induced lipid peroxidation of microsomal membranes is accompanied by a decrease of drug metabolizing activities and suggests that disruption of membranes by lipid peroxidation may lead to an inactivation of membrane-bound enzymes (40, 41, 42, 43). For example, Yukawa and Nakazawa (44) reported that irradiation of liver microsomes caused lipid peroxidation and a decrease in activities of the drug-metabolizing enzyme system through damage to cytochrome P-450. Addition of radical scavengers to microsomal suspensions protected both the radiation-induced lipid peroxidation and the decrease in the enzyme activities. Yukawa and coworkers then proceeded to examine the effect of radiation on the drug metabolizing enzyme system of microsomes reconstituted with liposomes of microsomal phospholipids, NADPH-cytochrome P-450 reductase and cytochrome P-450 (45). The reconstituted system of non-irradiated enzymes with irradiated liposomes showed a low activity of hexobarbital hydroxylation, whereas irradiated enzymes combined with non-irradiated liposomes exhibited an activity equal to that of unirradiated controls. These studies suggest that radiation-induced damage of membrane-bound enzymes might be caused as a result of conformational changes in the hydrophobic domain of the membrane lipid bilayer induced by lipid peroxidation.

The Participation of Radicals and Peroxides in Eicosanoid Synthesis

The synthesis of prostaglandins (PGs) involves hydroperoxy and free oxygen radical intermediates. Biosynthesis of PGs begins when arachidonic acid is released from lipid stores of the cell and is catalyzed by fatty acid cyclooxygenase, a constituent membrane bound enzyme of the PGH_2 synthetase complex, which converts the released arachidonate to 15-hydroperoxy-9,11-endoperoxide (46, 47, 48, 49, 50). The first step in the reaction sequence is the elimination of a hydrogen atom from the C-13 position of the precursor arachidonate, leaving a free radical intermediate. Thereafter, a peroxy radical is added to positions 9 and 11, an 8,12 carbon-carbon bond is formed, subsequent to the isomerization of the 11,12 double bond to 12,13, and another peroxy radical is added to position 15, resulting also in the isomerization of the 12,13 double bond to 13,14. The first product of this reaction sequence is the 15-hydroperoxy-9,11-endoperoxide, PGG_2. A peroxidase, which is also part of the PGH_2 synthetase complex, then catalyzes the reduction of the 15-hydroperoxy group on PGG_2 to a hydroxyl group, forming the 15-hydroxy-9,11-endoperoxide, PGH_2 (51, 52, 53). PGH_2 is then converted into the various PG products (PGE_2, PGF_{2a}, PGI_2, TXA_2, etc.) by the terminal enzymes in the biosynthetic pathway.

During the conversion of arachidonate into PGH_2, PGH_2 synthetase undergoes an irreversible self-deactivation (54). Using microsomes from ram seminal vesicles, Egan et al. (51) demonstrated that PGH_2 synthetase-cyclooxygenase was irreversibly self-deactivated due to the natural reduction of the hydroperoxide at carbon 15 of PGG_2 to the hydroxyl on PGH_2. During this reduction, radicals were detected. Both phenol and methional, scavengers of oxygen-centered radicals, promoted the formation of PGH_2 from PGG_2 and increased the initial rate and the extent of reaction prior to deactivation of the PGH_2 synthetase-cyclooxygenase. These workers proposed that during the conversion of PGG_2 to PGH_2, oxygen-centered radicals were generated which occasionally attacked the PGH_2 synthetase-cyclooxygenase, thereby deactivating it. Phenol and methional acted to protect the enzyme by trapping any radicals before they attacked the enzyme. These radicals produced by the reaction were also shown to deactivate the PGH_2 synthetase-peroxidase itself (52). PGI_2 synthetase activity in ram seminal vesicle microsomes has also been shown to become irreversibly deactivated by an oxidizing species generated during the PGH_2 synthetase-peroxidase-catalyzed reduction of PGG_2 to PGH_2 (55).

Despite the self-destructive nature of its catalytic activity, PGH_2 synthetase requires a certain ambient level of peroxides to function. The PGH_2 synthetase-cyclooxygenase product, PGG_2, has been shown to function as an activator, feeding back to stimulate further PGH_2 synthetase activity (56). A variety of other peroxides have been found to stimulate prostaglandin synthesis. Lipid hydroperoxides, formed by lipoxygenase reactions, are extremely potent activators of PGH_2 synthetase activity (56, 57). Hydrogen peroxide itself has also been shown to stimulate PGH_2 synthetase activity (58, 59). Work conducted in our own laboratory has demonstrated that H_2O_2 and t-butyl hydroperoxide increased or decreased prostaglandin synthesis in cultured human embryo lung fibroblasts, endothelial cells and corneal fibroblasts, depending on peroxide and radical trapper concentrations (60, 61, 62). In fact, PGH_2 synthetase activity is inhibited by conditions in which peroxides are not allowed to form (57, 63). On the other hand, excessive generation of peroxides and subsequent oxygen radicals inactivates PGH_2 synthetase but has no effect on PGE isomerase (64). Deby and Deby-Dupont also demonstrated that free radicals can either increase or decrease PG production from arachidonate (65). For example, hydrogen peroxide (at 10^{-4} to 10^{-5}M concentrations) and the hypozanthine-zanthine oxidase system (at 10^{-4} to 10^{-5}M concentrations) increased PG production from precursor arachidonate, whereas superoxide dismutase inhibited PG synthesis at low doses but stimulated PG synthesis at high doses. The balance between the formation and removal of cellular peroxides and radicals thus sets a peroxide tone which participates in the regulation of prostaglandin production.

The involvement of free radicals in the transformation of arachidonate to PGH_2 and the requirements of the PGH_2 synthetase complex for peroxides makes the biosynthesis of PGs susceptible to the cellular peroxide tone. Obviously, an agent such as radiation, which has the ability to alter the cellular levels of free radicals and peroxides, possesses the potential to influence PG biosynthesis. In fact, several studies suggest that free radicals induced by radiation can alter eicosanoid biosynthesis. For example, Donlon et al (66) demonstrated that pre-irradiation administration of the radioprotectant drug WR-2721, which is a free radical scavenger (67), to rats resulted in a significant reduction in radiation-induced increases in excretion rates of PGE, PGF_{2a} and TXB_2. A study in our laboratory showed that the presence of 4-hydroxy-2,2,6,6-tetramethylpiperidinoxy radical, an oxygen radical trapper, during gamma irradiation of endothelial cells in culture potentiated the radiation-induced increase in the synthesis of PGI_2 (68).

Lipid Peroxidation and Phospholipase A_2 Activity

Several studies have demonstrated that gamma radiation stimulates PLA_2 activity. Treatment of cultured bovine pulmonary artery endothelial cultures with radiation doses ranging from 1500 to 3000 rads stimulated the release of ^{3}H-arachidonate from prelabeled cellular lipid stores (69, 70). Indirect evidence of radiation-induced phospholipase A_2 stimulation is provided by studies examining the incorporation of ^{3}H-acetate into platelet activating factor (PAF). PAF synthesis involves the hydrolysis of 1-O-alkyl-2-acetyl-sn-glyceryl-3-phosphorylcholine from membrane phospholipids affected by phospholipase A_2 followed by the acetylation of 2-lyso PAF at position 2 and thus serves as a measure of phospholipase A_2 activity (71, 72). Results from our laboratory demonstrated that the exposure of endothelial cells to gamma radiation increased the intracellular levels of labeled PAF (68).

The activity of phospholipase A_2 appears to be affected by the generation of lipid peroxides. For example, studies conducted by Fujimoto and coworkers (74) demonstrated that lipid peroxidation induced by ascorbic acid and Fe^{+2} enhanced the release of arachidonate in kidney medulla slices. Another study by Ogawa and coworkers suggest that changes in membrane structure brought about by lipid peroxidation increases the action of PLA_2 (75). Also, the inhibition or the absence of PLA_2 activity has been found to reduce the activities of glutathione transferase and peroxidase (76, 77, 78, 79). Thus, the activation of phospholipase A_2 may function as an important regulator of the cellular lipid peroxides and subsequent generation of oxygen radicals. Indeed, exposure of lung fibroblasts to hydroperoxides, perhaps at levels above those affecting the action of PGH_2-cyclooxygenase, increases the release of arachidonate from cellular lipid stores (60, 62). Radiation may therefore be stimulating phospholipase A_2 activity by inducing the lipid peroxidation of cell membrane lipids.

Ultraviolet Radiation and Prostaglandin Synthesis

Ultraviolet radiation appears to influence the activity of both PGH_2 synthetase and phospholipase A_2. Ziboh et al (80) examined the in vivo release of arachidonate following UV-irradiation of guinea pigs at various intervals, and the in vitro effect of UV-irradiation on phospholipase A_2 activity. In the in vivo studies, maximum release of arachidonate was attained within the first 30 minutes, remained increased through 48 hours and finally decreased to normal levels 72 hours after irradiation. Phospholipase A_2 activity was assayed by incubating the high speed particulate fraction prepared from guinea pig skin with ^{14}C-labeled lecithin and cofactors for 3 minutes under UVB-irradiation. UV-irradiation was found to increase phospholipase A_2 activity by more than 4 fold. Lord et al (81) demonstrated that UV-irradiation (at 254 and 360 nm) of microsomes prepared from fetal calf skin enhanced PG synthesis from arachidonate. These authors postulate that UV radiation is involved in the initiation of a free-radical fatty acid intermediate from arachidonate which then serves as a reactive substrate for the attack of O_2. This free-radical oxygen complex is subsequently transformed by the PGH_2 synthetase enzyme into hydroperoxide and cyclic endoperoxide intermediates.

Evidence suggests that UV radiation can indeed induce the formation of radicals and lipid peroxides. The absorption of light has the ability to produce free radicals (23, 82). Pathak and Stratton (83) detected the production of free radicals in the skin after exposure to UV radiation. Meffert et al (84) showed that UV light caused lipid peroxidation in human skin and that the products increased with increasing exposure to UV light or sunlight. Mandal and Chatterjee (85) found that UV radiation and sunlight caused lipid peroxidation in the liposomal and erythrocyte membrane (as detected by the measurement of the oxidation index, A_{233}/A_{215}, and the amount of malondialdehyde formed). The oxidation index and the formation of malondialdehyde increased linearly with increasing dose of radiation and depended significantly on the dose rate. These findings thus suggest that UV radiation, like ionizing radiation, alters PG biosynthesis by generating free radicals and lipid peroxidation.

Conclusions

Cellular peroxide and oxygen radical concentrations appear to be important determinants in the action of endoperoxide synthetase and possibly phospholipase A_2. Ionizing and ultraviolet radiation have the ability to alter the peroxide levels of cells and tissues. Evidence suggests that radiation affects both phospholipase A_2 activity, and therefore the release of precursor arachidonic acid from cellular phospholipid stores, and the activity of the PG biosynthetic enzymes (ie. PGH_2 synthetase). The possible radiation-induced mechanisms for these alterations include the increased lipid peroxidation of cell membrane lipids and/or the formation of oxygen-derived free radicals. Whether or not radiation stimulates or inhibits eicosanoid synthesis may well depend on the concentrations of peroxides and radicals generated by irradiation and the ability of the tissue to handle these compounds.

References

1. Edwards, J.C., Chapman, D., Cramp, W.A. and Yatvin, M.B. Progr. Biophys. Mol. Biol. 43:71-93, 1984.
2. Anbar, M. and Neta, P. Int. J. Appl. Radiat. Isot. 18:493-523, 1967.
3. Petkau, A. Can. J. Chem. 49:1187-1195, 1971.
4. Petkau, A. and Chelack, W.S. Int. J. Radiat. Biol. 25:321-328, 1974.
5. Petkau, A. and Chelack, W.S. Int. J. Radiat. Biol. 26:421-426, 1974.
6. Behar, D., Czapski, G., Dorfman, L.M., Rabani, J. and Schwarz, H.A.J. Phys. Chem. 74:3209-3213, 1970.
7. Klug, D., Rabani, J. and Fridovich, I. J. Biol. Chem. 247:4839-4842, 1972.
8. Goda, K., Kimura, T., Thayer, A.L., Kees, K. and Schaap, A.P. Biochem. Biophys. Res. Commun. 58:660-666, 1974.
9. Rawls, H.R. and Van Santen, P.J. J. Am. Oil Chem. Soc. 47:121-125, 1970.
10. Rawls, H.R. and Van Santen, P.J. Ann. N.Y. Acad. Sci. 171:135-137, 1970.
11. Halliwell, B. New Phytol. 73:1075-1086, 1974.
12. McCord, J.M. and Fridovich, I. J. Biol. Chem. 244:6049-6055, 1969.
13. Pederson, T.C. and Aust, S.D. Biochem. Biophys. Res. Commun. 52:1071-1078, 1973.
14. Debey, P. and Balny, C. Biochimie 55:329-332, 1973.
15. Tyler, D.D. FEBS Lett. 51:180-183, 1975.
16. Petkau, A. and Chelack, W.S. Biochim. Biophys. Acta 433:445-456, 1976.
17. Inouye, B., Aono, K., Iida, S. and Utsumi, K. Physiol. Chem. & Physics 11:151-160, 1979.
18. Haber, F. and Weiss, J. Proc. R. Soc. Ser. A. 147:332-351, 1934.
19. Singh, A. and Singh, H. Prog. Biophys. Molec. Biol. 39:69-107, 1982.
20. Leyko, W. and Bartosz, G. Int. J. Radiat. Biol. 49:743-770, 1986.
21. Myers, D.K. In:Advances in Biological and Medical Physics (Eds. J.H. Lawrence and J.W. Gofman), Academic Press, New York, 1970, Vol. 13, pp. 219-234.
22. Wilbur, K.M., Bernhein, F. and Shapiro, O.W. Arch. Biochem 24:305-313, 1949.
23. Ingold, K.U. Chem. Rev. 61:563-589, 1961.
24. Dahle, L.K., Hill, E.G. and Holman, R.T. Arch. Biochem. Biophys. 98:253-261, 1962.
25. Pryor, W.A. Fed. Proc. 32:1862-1869, 1973.
26. Nakazawa, T., Nagatsuka, S. and Sakurai, T. Int. J. Radiat. Biol. 40:365-373, 1981.

27. Mooibroek, J., Trieling, W.B. and Konings, A.W.T. Int. J. Radiat. Biol. 42:601-609, 1982.
28. O'Connell, M.J. and Garner, A. Int. J. Radiat. Biol. 44:615-625, 1983.
29. Konings, A.W.T. Int. J. Radiat. Biol. 40:441-444, 1981.
30. Konings, A.W.T., Damen, J. and Trieling, W.B. Int. J. Radiat. Biol. 35:343-350, 1979.
31. Wills, E.D. and Wilkinson, A.E. Radiat. Res. 31:732-747, 1967.
32. Barber, D.J.W. and Thomas, J.K. Radiat. Res. 74:51-65, 1978.
33. Purohit, S.C., Bisby, R.H. and Cundall, R.B. Int. J. Radiat. Biol. 38:C147-158, 1980.
34. Purohit, S.C., Bisby, R.H. and Cundall, R.B. Int. J. Radiat. Biol. 38:159-166, 1980.
35. Nakazawa, T., Nagatsuka, S. Int. J. Radiat. Biol. 38:537-544, 1980.
36. Cataldi de Flombaum, M.A. and Stoppani, A.O.M. Biochem. International 12:785-793, 1986.
37. Das, D.K. and Neogi, A. Clin. Physiol. Biochem. 2:32-38, 1984.
38. Burkhardt, H., Scwingel, M., Menninger, H., McCartney, H.W. and Tschesche, H. Arthritis and Rheumatism 29:379-387, 1986.
39. Ortiz, X., Ragsdale, C. and Test, S.T. Science 227:747-749, 1985.
40. Wills, E.D. Biochem. J. 123:983-991, 1971.
41. Kamataki, T. and Kitagawa, H. Biochem. Pharmacol. 22:3199-3207, 1973.
42. Koster, J.F. and Slee, R.G. Biochem. Biophys. Acta 620:489-499, 1980.
43. Wills, E.D. and Wilkinson, A.E. Int. J. Radiat. Biol. 17:229-236, 1970.
44. Yukawa, O. and Nakazawa, T. Int. J. Radiat. Biol. 37:621-631, 1980.
45. Yukawa, O., Nagatsuka, S. and Nakazawa, T. Int. J. Radiat. Biol. 43:391-398, 1983.
46. Hemler, M. and Lands, W.E.M. J. Biol. Chem. 251:5575-5579, 1976.
47. Van der Ouderaa, F.J., Buytenhek, F.J., Nugteren, D.H. and Van Dorp, D.A. Biochim. Biophys. Acta 487:315-331, 1977.
48. Hamberg, M. and Samuelsson, B. Biochem. Biophys. Res. Commun. 61:942-949, 1974.
49. Hamberg, M. and Samuelsson, B. Proc. Natl. Acad. Sci. (USA) 71:3400-3404, 1974.
50. Nugteren, D.H. and Hazelhof, E. Biochim. Biophys. Acta 326:448-461, 1973.
51. Egan, R.W., Paxton, J. and Kuehl, F.A., Jr. J. Biol. Chem. 251:7329-7335, 1976.
52. Egan, R.W., Gale, P.H. and Kuehl, F.A., Jr. J. Biol. Chem. 254:3295-3302, 1979.
53. Miyamoto, T., Oginn, N., Yamamoto, S. and Hayaishi, O. J. Biol. Chem. 251:2629-2636, 1976.
54. Smith, W.L. and Lands, W.E.M. Biochem. 11:3276-3285, 1972.
55. Ham, E.A., Egan, R.W. Soderman, D.D., Gale, P.H. and Kuehl, F.A., Jr. J. Biol. Chem. 254:2191-2194, 1979.
56. Hemler, M.E., Lands, W.E.M. and Graff, G. Biochem. Biophys. Res. Commun. 85:1325-1331, 1978.
57. Hemler, M.E., Cook, H.W. and Lands, W.E.M. Arch. Biochem. Biophys. 193:340-345, 1979.

58. Panganamala, R.V., Sharma, H.M., Sprecher, H., Greer, J.C. and Cornwell, D.G. Prostglandins 8:3-11, 1974.
59. Morse, D.E., Duncan, H., Hooker, N. and Morse, A. Science 196:298-300, 1977.
60. Polgar, P. and Taylor, L. Prostaglandins 19:693-700, 1980.
61. Taylor, L., Menconi, M., Leibowitz, H. and Polgar, P. Invest. Opthalmol. & Vis. Sci. 23:378-382, 1982.
62. Taylor, L., Menconi, M.J. and Polgar, P. J. Biol. Chem. 258:6855-6857, 1983.
63. Hemler, M. E. and Lands, W.E.M. J. Biol. Chem. 255:6253-6261, 1980.
64. Fujimoto, Y, Tanioka, H., Keshi, I. and Fujita, T. Biochem. J. 212:167-171, 1983.
65. Deby, C. and Deby-Dupont, G. In:Biological and Clinical Aspects of Superoxide and Superoxide Dismutase (Eds. W.H. Bannister and J.V. Bannister), Proc. Fed. Eur. Biochem. Soc. Symposium No. 62, 1982, pp. 84-97.
66. Donlon, J., Steel, L., Helgeson, E.A., Wolfe, W.W. and Catravas, G.N. Int. J. Radiat. Biol. 47:205-212, 1985.
67. Fowler, J.F. Int. J. Radiat. Oncol. Biol. Phys. 11:665-674, 1985.
68. Menconi, M., Taylor, L. and Polgar, P. In:Prostaglandin and Lipid Metabolism in Radiation Injury (Eds. T.L. Walden, Jr. and H.N. Hughes), 1987, in press.
69. Friedman, M., Saunders, D.S., Madden, M.C., Chaney, E.L. and Kwock, L. Radiat. Res. 106:171-181, 1986.
70. Hahn, G.L., Menconi, M.J., Cahill, M. and Polgar, P. Prostaglandins 25:783-791, 1983.
71. Camussi, G., Aglietta, M., Malavasi, F., Tetta, C., Piacibello, W., Sanavio, F. and Bussolino, F. J. Immunol. 131:2397-2403, 1983.
72. Prescott, S.M., Zimmerman, G.A. and McIntyre, T.M. Proc. Natl. Acad. Sci. (USA) 81:3534-3538, 1984.
74. Fujimoto, Y. and Fujita, T. Biochim. Biophys. Acta 710:82-86, 1982.
75. Ogawa, Y., Kondo, T., Sugiyama, S., Ogawa, K., Satake, T. and Ozawa, T. Cancer Res. 47:1239-1243, 1987.
76. Hong, K., Meyer, D.J., Belin, J. and Ketterer, B. Biochem. J. 220:243-252, 1984.
77. Sevanian, A., Muakkassah-Kelly, S.F. and Montestruque, S. Archiv. Biochim. Biophys. 441-452, 1983.
78. Shewfelt, R. and Hultin, H.O. Biochim. Biophys. Acta 751:432-438, 1983.
79. Grossmann, A. and Wendel, A. European J. Biochem. 135:549-552, 1983.
80. Ziboh, V.A., Lord, J.T., Uematsu, S. and Blick, G. J. Invest. Dermatol. 70:211, 1978.
81. Lord, J.T., Ziboh, V.A., Poitier, J., Legget, G. and Penneys, N.S. Br. J. Dermatol. 95:397-406, 1976.
82. Orlov, S.N., Malkov, Y.A., Rebrov, V.G. and Danilov, V.S. Biophysics (USSR) 21:278-282, 1976.
83. Pathak, M.A. and Stratton, K. Arch. Biochem. Biophys. 123:468-476, 1968.

84. Meffert, H., Diezel, W. and Sonnichsen, N. Experientia 32:1397-1398, 1976.
85. Mandal, T.K. and Chatterjee, S.N. Radiat. Res. 83:290-302, 1980.

11

RADIOBIOLOGY OF BLOOD VESSELS

Eliot M. Rosen, M.D., Ph.D.
David W. Vinter, Ph.D.
Itzhak D. Goldberg, M.D.

Why should the cancer specialist be interested in the radiobiology of blood vessels? This chapter will attempt to answer this question, review the current state of knowledge of vascular radiation biology and suggest directions for future investigations. Radiation effects on normal tissues are divided into two categories based on time of appearance. Acute effects occur during a course of fractionated radiation therapy or shortly thereafter. Late effects occur more than two months after completion of radiation and may appear many years later. Acute radiation damage to normal tissues can be understood in terms of cell proliferation kinetics (1). It results from an imbalance between cell killing and tissue regeneration. The latter may occur by recruitment of quiescent "stem" cells into the cell cycle; shortening of the cycle time of proliferating cells; or repopulation of the tissue with cells from outside of the irradiated field. The timing and severity of the acute effects depend upon: (1) transit time for maturing cells within the tissue compartment; (2) dose fraction size; and (3) protraction (the total time in which the radiation is delivered). Examples of normal tissues susceptible to acute damage (and the corresponding effects) are: bone marrow (cytopenias); oropharyngeal mucosa (mucositis); and skin (desquamation).

While acute effects can be modified during irradiation by reducing the daily dose fraction or instituting a treatment break, late effects are not conveniently modifiable. Late radiation damage is determined primarily by total dose, dose fraction size, and irradiation volume, and is truly the dose-

P. Polgar (ed.), EICOSANOIDS AND RADIATION.

limiting factor in clinical radiotherapy. Late damage does not correlate well with the degree of acute damage. This lesson was learned by clinicians who were able to deliver large doses of radiation (8000 cGy or more) to breast and head and neck tumors with little or no acute skin or mucosal reaction by manipulation of the fractionation schedule (2). Despite the virtual absence of acute reactions, severe skin atrophy, telangiectasias, and tissue fibrosis ensued.

Some investigators have employed a modification of the cell kinetic model discussed above to explain late organ damage. According to this scheme, late-reacting tissues contain slowly cycling cell populations in which cell death may be delayed for many months. Ultimately, depletion of these parenchymal "stem" cells beyond the reserve functional capacity of the tissue results in expression of late damage (3,4). Late damage to tissues in which the major cell type is non-proliferating (such as central nervous system and skeletal muscle) is postulated to result from depletion of supporting or stromal cells (5,6). This hypothesis does not account for the possibility that organ dysfunction might not be directly related to cell sterilization. Dose-effect curves for organ function and clonogenic cell survival may differ substantially, and both may affect radiation tolerance dose (7,8).

Another hypothesis attributes late tissue effects to damage to small blood vessels (capillaries, arterioles and venules) (9,10). This hypothesis is based upon histopathologic evidence of breakdown and obliteration of small vessels in association with tissue damage, suggesting disruption of blood supply secondary to radiation-induced vascular injury as the etiologic event. Modifications of the vascular hypothesis state that tissue damage results from initial functional rather than structural vascular alterations:

(1) radiation → increased vascular permeability → accumulation of clotting factors in the interstitium → fibrin formation → fibrogenesis → increased diffusion barrier.

(2) radiation → decreased blood flow → ischemic tissue damage.

If the vascular hypothesis is used as a generic explanation for late effects in different organs, we must ask why various organs show disparate radiation tolerance doses (eg., kidneys tolerate approximately 2000 cGy, lungs 1500, spinal cord and small bowel 4500-5000, and rectum 6000, when delivered in 180-200 cGy fractions). Unless we postulate differences in the intrinsic radiosensitivities of the various capillary beds, other explanations must be sought. For example, the degree of tissue damage and time of its onset may depend upon the pattern of vasculature and its redundancy; the fraction of cycling endothelial cells; the sensitivity of the parenchymal cells to alterations in blood supply; the reserve capacity of the tissue; and the relationship between tissue function and damage to the functioning units.

To summarize, the exact cause of late tissue damage, the critical dose-limiting factor in clinical radiation therapy, is unknown. It may involve damage to parenchymal cells (3,4); blood vessels (9,10), supporting cells (6), connective tissue stroma (11), and more likely, some combination of the above. It is likely that radiation injury to blood vessels plays a role in some forms of normal tissue damage. This is one reason that we should be interested in vascular radiation biology.

In addition to normal tissue damage, irradiation of blood vessels has other consequences. Radiation therapy to a tumor bed may result in a decreased blood supply due to damage to tumor vasculature (12,13). Alternatively, radiation can increase the blood supply by destruction of tumor parenchyma in excess of vascular damage or by stimulation of neovascularization (12). Hypoxic cells are more radioresistant than their normoxic counterparts. Clinical data as well as animal studies suggest an increased susceptibility to metastases in previously irradiated lung or liver (14). This phenomenon may be mediated by damage to small blood vessels allowing increased penetration and invasion by tumor cells (15,16). Therefore, effects of radiation on the microvasculature can influence local tumor control as well as rates of metastasis. This is another reason we should be interested in vascular radiobiology.

Vascular radiobiology can be approached at a number of levels: clinically observable effects on normal tissues and tumors; histopathologic features of damage to blood vessels; functional alterations, including changes in blood flow, permeability, and vascular metabolism; radiobiology of cultured vascular cells; and finally, subcellular and biochemical alterations. We will attempt to review and correlate the currently available radiobiologic data at these levels and to suggest possible avenues for future research.

Radiation Injury to Small Vessels and Late Tissue Damage. Supporters of the vascular hypothesis have proposed several possible target lesions: endarteritis of small arteries and arterioles; damage to capillary endothelium; and endothelialitis of arterioles and venules (9,10,17). The microvasculature is radioresponsive and shows significant morphologic and functional alterations even at relatively low doses. In tissues which show acute epithelial damage (e.g., skin, small intestine), the earliest vascular response is a hyperemia associated with increased permeability and edema (17). The onset of this response may be days or weeks depending upon radiation dose, fractionation, and animal species. This early response is felt to reflect a transient inflammatory process related to release of vasoactive mediators (e.g., histamine, prostaglandins) (18,19). A later vascular response, usually observed after 3-4 months or more, is characterized by capillary fibrosis and obliteration, endarteritis, and telangiectasias. This response may be associated with temporary or permanent reduction in blood flow.

The clearest histopathologic evidence of a vascular etiology for normal tissue damage is the response of the liver to radiation. Data from humans and animals indicate that the predominant early hepatic lesion is an endothelialitis of the hepatic sinusoids, followed by obliterative changes and thromboses in the centrilobular veins (10). The appearance of irradiated hepatic lobules is similar to that of hepatic veno-occlusive disease due to other causes (Budd-Chiari-like syndrome). Clinically, doses of 2500 to 3000 cGy in 10 fractions can cause a hepatitis-like syndrome within a few months of irradiation, and higher doses may result in death due to hepatic failure (20,21). In the rat model, late damage (hepatocyte loss and fibrosis) from irradiation of an exteriorized lobe of liver was preceded by vascular lesions (including obstruction of central veins), which started at 10 weeks (22).

Experimental evidence supporting the vascular hypothesis comes from a series of investigations of pig skin, considered a good model for human skin (9). These studies showed two waves of reaction after skin irradiation. The first peaked after 4-6 weeks and was characterized by bright erythema due to increased dermal blood flow, associated with epidermal reactions (desquamation). The second wave peaked at 10-14 weeks and was characterized by a dusky coloration indicative of ischemia and histopathologic evidence of arteriolar occlusion by endothelial cells in the deep dermal plexus. Regional blood flow was reduced at this time, and the degree of reduction correlated with the degree of late skin atrophy. The vascular flow abnormality returned to normal over the period from 12 to 26 weeks. Despite apparently normal blood flow at 1-2 years, raising the irradiated skin segment as a pedicle flap resulted in necrosis, suggesting impaired capacity of the vasculature to respond to additional stress.

Van der Kogel (6) studied radiation-induced myelopathy in the rat spinal cord and concluded that two different pathologic mechanisms are involved. An early syndrome, encountered after high doses, was characterized by a short latent period, rapid symptomatic progression, and predominant lesions of white matter necrosis. The oligodendrocyte was felt to be the critical target cell. A late syndrome occurred after lower doses and a longer latent period, was usually slowly progressive, and was associated with vascular lesions involving capillaries and venules. Similarly, damage to rat brain appeared to be mediated through glial cell depletion at high doses and vascular injury at lower doses (9,23).

To summarize, vascular injury probably contributes significantly to some forms of late normal tissue damage. The most important vascular lesions in these circumstances involve very small vessels: capillaries, arterioles, and venules.

Radiation Injury to Large Blood Vessels. Large veins and arteries are thought to be unusually resistant to radiation injury (24-26). Rubin estimates the dose of fractionated radiation necessary for 5% serious late complications to be 8000 cGy or higher (25). Irradiation of large arteries in animals results in little detectable effect at single doses less than 1500-2000 cGy (27-30).

Effects of radiation on large vessels of clinical significance have been reported in several groups of patients. Long-term survivors of Hodgkin's disease and other malignancies who received moderate doses of fractionated irradiation (4000-5000 cGy) to the mediastinum may be at risk for accelerated atherosclerosis in irradiated arteries. These patients appear to have an increased incidence of transient ischemic attacks and myocardial infarction, with angiographic evidence of narrowing and placque formation in irradiated vessels (31-34). Experimental studies in animals including pigeons, rabbits, rats and dogs have demonstrated accelerated development of atherosclerosis in the irradiated segments (29,30,35,36). Atherogenic diets may contribute to the development of lesions, which appear similar to those developing later in life in unirradiated animals. In a study of dog aortas treated with 50KVP x-rays, there was a suggestion that lower doses produced more severe lesions, when examined 1-11 months after irradiation (30). A relatively rare form of vascular damage is rupture of major arteries following radical surgery and high dose radiotherapy (37). Ruptures occurred at a median of 13 months after irradiation, and surgical complications were felt to be a major contributing factor. Another rare complication is abdominal aortic hypoplasia with renovascular hypertension after abdominal irradiation in early childhood (38).

Several investigators have reported histopathologic studies of radiation effects in large vessels in animals (11,28,39). Endothelial injury, with nuclear disruption, denudation, and patchy fibrin deposition, was noted within 48 hours after fractionated irradiation (4000 Roentgens in 10 days) to the dog femoral artery (40). Within 3 weeks, substantial regeneration of the luminal surface had occurred. Medial and adventitial damage appeared later and were more slowly progressive, with medial disorganization and fibrosis and adventitial hemorrhage and inflammation. The authors suggest that medial and adventitial changes may be longer lasting or permanent in comparison with endothelial changes, which are gradually repaired. In another study, single doses of x-rays were given to the mouse femoral artery and the radiation effects were scored quantitatively for up to 4 months (28). This study showed the most severe and long-lasting effects on the adventitia followed closely by the media. Intimal damage was felt to be minor. The authors of both studies concluded that intimal and medial-adventitial damage proceed through different mechanisms, the latter perhaps mediated in part by damage to adventitial capillaries and nerves or to the perivascular stroma (11).

Current concepts of atherogenesis have focused on endothelial injury resulting in (i) release of SMC growth stimulating factor(s); (ii) exposure of underlying SMC to blood-borne elements and products of hemostasis; and (iii) intimal proliferation of SMC as essential pathogenetic processes (40). We investigated the effects of irradiation on in vivo aortic smooth muscle cell (SMC) kinetics in the rat (41). Aortas were irradiated using 250 KVP x-rays with doses from 300 to 1500 cGy. Immediately following irradiation, aortas were balloon catheter denuded of endothelium to provide a stimulus for SMC proliferation. Utilizing an assay for specific activity of DNA (42), DNA synthesis 48 hours after irradiation was suppressed relative to control in a dose-dependent fashion (Figure 1). Electron microscopy of aortas which had received 1500 cGy 28 days earlier revealed proliferative intimal lesions, suggesting escape of SMC from the inhibitory effects of irradiation. Radiation injury may feed into a common pathway involving endothelial injury, intimal proliferation of SMC, and ultimately clinical evidence of atherosclerosis. As suggested by our data and the dog studies, there may be a balance between stimulation of SMC proliferation and initial SMC damage such that atherogenesis is inhibited at very high doses.

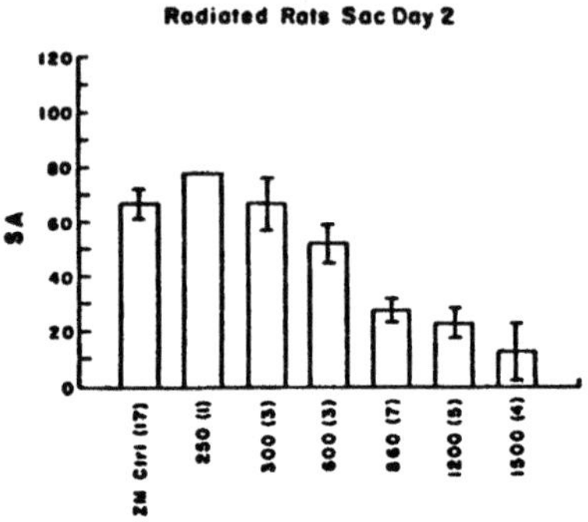

Figure 1. DNA specific activity (SA) (dpm per ug DNA) vs radiation dose (cGy) for rat thoracic aortic smooth muscle cells irradiated in vivo. Rats were irradiated with 250 KVP x-rays, balloon-denuded of endothelium immediately thereafter, and sacrificed 48 hr after balloon injury. One hr prior to sacrifice, they were given (3H)-thymidine (0.5uCi/gm weight) and the thoracic aorta was removed for SA measurement.

Radiobiology of Cultured Vascular Cells.

A. Clonogenic Survival Curves. The radiosensitivity of cultured large vessel endothelial cells (EC) and smooth muscle cells (SMC) has been studied utilizing *in vitro* clonogenic survival measured by colony formation assay as the end-point (15,43-50). Figures 2A and B show typical radiation dose-response curves from our studies of rat aortic SMC cultures (43-45). When the surviving fraction (relative to an unirradiated control) is plotted vs dose on semi-logarithmic coordinates, there is an initial shoulder followed by a straight-line region of exponential survival. The survival curve parameters typically calculated are: (1) extrapolation number (n) - obtained by back-extrapolation of the straight-line (exponential) portion of the curve to dose = 0; (2) Do - the incremental dose in the exponential region required to reduce survival by a factor of 1/e; and (3) quasi-threshold dose (Dq) - the dose obtained when the exponential portion of the curve is back-extrapolated to a survival fraction of unity: Dq=Do ln(n). The parameters Dq and n reflect the size of the shoulder, a measure of the cells' ability to accumulate sublethal radiation damage. Do is inversely related to the steepness of the slope of the survival curve: thus lower Do = steeper slope = more radiosensitive. Figures 2A and B demonstrate another feature of interest. The initial portion of the survival curve also has a negative slope, indicating cell damage even at the lowest doses. The steepness of the initial slope may be characterized by an "initial Do".

Exponentially growing cultures of early passage rat aortic medial SMC showed Do values of about 130 cGy (107-170); n about 6.5 (1-17); Dq about 210 cGy (30-352); and initial Do about 225-275 cGy (44). A split-dose experiment (600 cGy at time=0; incubate for variable time; then another 600 cGy) demonstrated maximal repair of sublethal radiation damage in about 4-6 hours (Figure 2C). Recovery factors (ratio of maximal survival fraction to that for 0 time between two doses) were approximately equal to the extrapolation numbers, suggesting a relationship between accumulation and repair sublethal damage. We measured the survival properties of confluent multilayered SMC cultures by trypsinizing and plating cells for colony formation assays immediately after radiation (45). These cultures had similar survival parameters compared with sparse proliferating cultures or were slightly more radioresistant (Figures 3A,B).

Potentially lethal radiation damage may or may not be expressed depending upon post-irradiation conditions (51). Potentially lethal damage repair (PLDR) occurs under conditions in which cell proliferation is slow, as in confluent plateau-phase cultures. PLDR in multilayered SMC cultures was assayed by incubation for up to 24 hours prior to plating for colony survival. PLDR was nearly maximal in 8 hours (Figure 3C) and occurred in a dose-dependent fashion. Repair factors (ratio of survival fraction for 24 hour incubation to that for immediate plating) for a typical culture were: 1.3 at 500 cGy; 5.4 at 1000 cGy; and 17 at 1500 cGy. These findings suggest that during a clinical course of fractionated radiation (200 cGy per day), PLDR in SMC is probably insignificant. However, after large single doses, as utilized in patients receiving

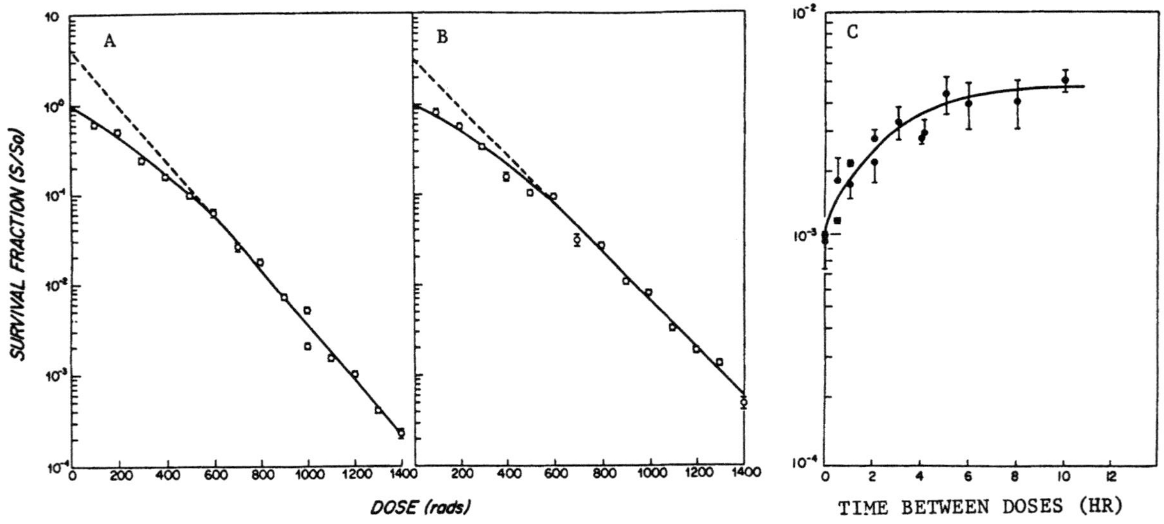

Figure 2. Panels A and B: Radiation dose-response curves for two different strains of cultured rat thoracic aortic smooth muscle cells. Panel A shows data for strain 231T at the 11th population doubling: Do=143 cGy, n=3.92, Dq=195 cGy. Panel B shows data for strain 232 at the 9th doubling: Do=162 cGy, n=3.21, Dq=189 cGy.
Panel C: Split dose radiation survival curve for smooth muscle cell strain 141T at the 9th population doubling. Cultures received 600 cGy at time=0 and another 600 cGy at varying times from 0-10 hours. The recovery factor was about 4.5. Data points represent mean ± CEM for triplicate determinations.

intraoperative radiation therapy (1200-3000 cGy) and in animal studies, PLDR may contribute to the radioresistance of large blood vessels.

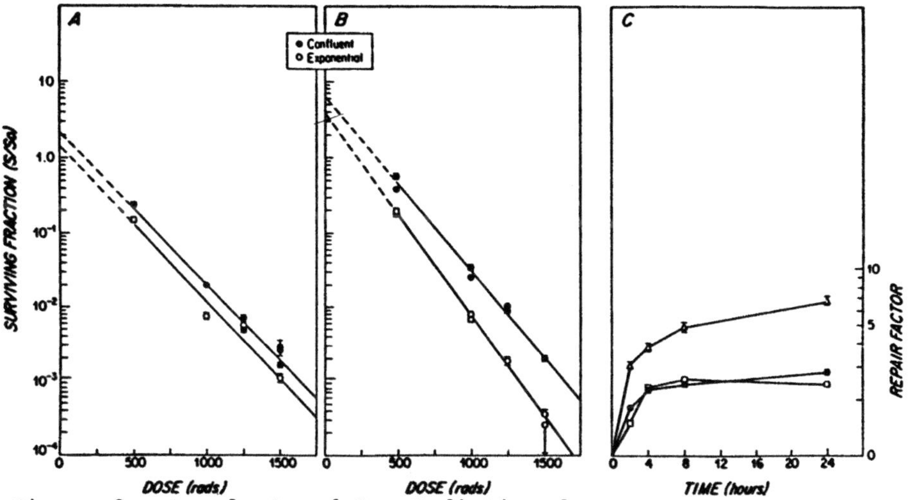

Figure 3. Panels A and B: Radiation dose-response curves comparing confluent and exponentially growing cultures of rat aortic smooth muscle cells. Exponential cultures were derived from control (0 dose) confluent cultures and were irradiated 24 hours after plating. Panel A shows data for cell strain 231T and Panel B shows data for strain 223T. Panel C: Time course of potentially lethal damage repair for three strains of smooth muscle cells (open circles, 223T; closed circles, 231T, squares 232T). Cultures received 1000 cGy at time=0 and were incubated in fresh medium for varying times up to 24 hours prior to subculture for colony formation assays. Data points are mean ± SEM for triplicate measurements.

Animal age (3-23 months) and site (thoracic vs abdominal aorta) did not significantly influence the radiosensitivity of early passage SMC cultures (44). However, there was a trend toward decreased shoulder (n), increased resistance (Do), and increased plating efficiency with increasing culture age. The net effects were that cultures became "heartier" and more radioresistant as they adapted to cell culture. From another point of view, these findings might suggest a larger shoulder *in vivo* than in early passage cultures, reflecting increased survival at low doses.

In the upper half of Table 1 are summarized radiation survival parameters for cultured large vessel EC and SMC by clonogenic survival assays. The radiosensitivity of EC was

Table 1. Radiation dose-effect parameters for endothelial and smooth muscle cells derived from large vessels

	Cell Type[1]	Culture Age	Assay End-Point	Growth State	Dose-Effect Parameters[2] Do(cGy)	n	Dq(cGy)	Reference
1.	Rat aortic SMC	7-12 Doublings	Clonogenic survival	Exponential	130±20	6.5±4	210	Rosen (44)
				Confluent	Do,n same or slightly higher			Rosen (45)
2.	Rat aortic SMC	13th-18th Pass	Clonogenic survival	Confluent	160	~3.7	~210	Heisel (15)
3.	Rabbit aortic EC	Established cell line	Clonogenic survival	Exponential	120	7	230	Martin (46)
				Confluent	120	5	193	
4.	Bovine aortic EC	5th-10th Pass	Clonogenic survival	Confluent	140	~2.4	~120	Heisel (15)
5.	Bovine pulmonary artery EC	Early passage	Clonogenic survival	Cell suspension	161±35	1.6±0.2	71±13	Kwock (49)
6.	Bovine aortic EC	7th-15th Pass	Clonogenic survival	Exponential	101	1.9	65	Rhee (50)
1.	Human umbilical vein EC	8th Pass	Inhibition of growth 4 days after radiation	Exponential	160	(~1)	(~0)	DeGowin (50)
2.	Aortic EC	Early passage	Inhibition of growth 7 days after radiation	Exponential	40	15	110	Griem (53)
	medial SMC				220	(~1)	(~0)	
	adventitial cells from rabbit				220	(~1)	(~0)	

[1]EC = endothelial cells; SMC = smooth muscle cells
[2]See text for definitions of survival parameters

generally similar to that for SMC, with Do values of 100-160 cGy. These values are similar to Do's measured in vitro in a variety of mammalian cell lines and in vivo for epithelial tissues in rodents (53,55). Three studies of bovine aortic and pulmonary artery EC showed smaller shoulders (lower Dq) as compared with an established line of rabbit aortic EC and with rat aortic SMC. If a small shoulder is characteristic of EC, it may suggest increased susceptibility to small daily dose fractions (200 cGy) used in clinical radiotherapy. EC repaired sublethal damage between two doses of radiation (46,50), and confluent monolayers exhibited PLDR (repair factor of 2 at 750 cGy) (46). Reducing the dose rate from 110 to 3 cGy/min resulted in an increase in Do from 102 to 147 cGy, consistent with the observation that cells repaired sublethal damage (50).

B. Inhibition of DNA Synthesis and Cell Replication. In the lower half of Table 1 are radiation dose-effect data utilizing inhibition of cell growth as the end-point. These Do values may reflect factors other than cell survival, including mitotic delay, cell detachment, cell cycle alterations, and ability of lethally irradiated cells to undergo a few divisions before dying (56). These data suggest large vessel EC are more susceptible to the anti-proliferative effects of radiation than are SMC.

This conclusion is supported by the data of Johnson and colleagues (57) who compared the Cs-137 gamma radiation response (250-2000 cGy) of confluent cultures of bovine aortic EC and SMC. While the attachment efficiency post radiation was similar in both cell types, growth inhibition of cells dissociated and plated 72 hours after irradiation was up to ten-fold greater in EC than in SMC, when examined 4 days after attachment. In contrast with SMC cultures, which showed little or no morphologic effect, EC cultures showed progressive cell loss over the 3 days following radiation (70% at 1000 cGy) in association with ultrastructural lesions (increase in lysosomes; membrane defects; irregular chromatin aggregation).

Studies of the longer-term effects of radiation on proliferating human umbilical vein EC cultures revealed that a dose of 750 cGy abolished net cell growth for at least 3 weeks

(52). DNA synthesis was suppressed for at least 2 weeks after 870 cGy (58). Similarly, EC from confluent cultures given 870 cGy failed to grow in 2 weeks after passage to a lower cell density. Human fibroblasts showed less inhibition of growth as compared with EC.

Early cell cycle events were monitored in confluent bovine aortic EC cultures exposed to Co-60 gamma radiation (59). The earliest response was a reduction in (3H)-thymidine incorporation per cell, occurring 2 hours after 500 cGy. This was followed by cell cycle redistribution (increased proportion of S-phase cells at 4 hours and decreased S-phase at 24 hours), cell detachment, and LDH release. Reduction of autoradiographic labelling index was the least sensitive end-point, occurring 24 hours after 500 cGy. The early increase in S-phase might be due to S-phase prolongation or arrest, or to selective loss of non S-phase cells. The later decrease might be caused by progression of a partially synchronized EC population.

If large vessel EC cultures are susceptible to cell killing, inhibition of proliferation, and structural damage, how is *in vivo* endothelial damage repaired? The radioresistance of large arteries and veins suggests that *in vivo* phenomena such as slow cell turnover and other complex vascular recovery mechanisms may be operative. Irradiation of the rat aorta with single large doses (up to 2000 cGy) did not result in an increase in the *in vivo* thymidine labelling index from 3-15 weeks after irradiation (50). The normal average adult aortic EC turnover rate is about 0.5% per day (about 10% in 3 weeks). Moreover, in areas of hemodynamic stress (near ostia and bifurcations), EC replication rates are 10% per day or higher (61). The *in vitro* data suggest that a dose of 2000 cGy should reduce the surviving fraction to less than 10^{-6} and delay proliferation by at least 3 weeks. It appears unlikely that proliferation by surviving EC could explain recovery of the endothelial lining.

Very efficient mechanisms of endothelial recovery exist in non-irradiated large vessels. Acute stresses such as hypertension and endotoxemia result in a 5-fold increase in EC turnover without appearance of denudation (61). Small denuding injuries are repaired rapidly by cell migration from the edge of the wound without replication (62). Large denuding injuries are repaired by migration of EC from outside of the wound with subsequent replication (63). Gentle injury confined to the intima is repaired in only 2-3 days. More severe damage may take several weeks or up to 1-2 months to repair. Like denudation, endothelial coverage of vascular grafts occurs rapidly and efficiently (64). Circulating EC have been documented by a number of investigators (65,66), and may also play a role in endothelial regeneration. Fischer (67) transplanted irradiated male aortic segments into female rats and found EC with Barr bodies in the irradiated segments.

These data suggest repopulation by unirradiated cells as a mechanism of endothelial recovery from radiation damage. To prove this point, careful studies of EC repopulation kinetics are required. For example, if recovery is due to migration of unirradiated EC into the radiation field from both edges and subsequent replication, (3H)-thymidine labelling studies should show advancing borders of DNA-synthesizing cells. There should also be an increase in DNA-synthesis in EC outside of the radiation field to compensate for cells lost by migration. In addition, both _in vitro_ (57) and _in vivo_ (68) studies suggest that irradiated EC can hypertrophy to cover a larger surface area, similar to the process that occurs during _in vitro_ (69) and _in vivo_ (70) aging.

In contrast to the endothelium, recovery by repopulation from distant sites is unlikely for SMC in the compact environment of the vascular media, in which individual SMC are tightly enclosed by other SMC and embedded in connective tissue matrix. Here, a combination of very slow turnover and continued ability of SMC to produce connective tissue matrix after irradiation might account for the ability of large arteries to remain structurally intact.

C. _Effects of Radiation on Vascular Cell Function_. The endothelium is a pluripotent tissue with multiple physiologic functions: (1) maintenance of an intact non-thrombogenic vascular lining; (2) active transport of special molecules between blood and underlying tissues; (3) interactions with the coagulation, fibrinolytic, and complement systems; (4)

metabolism of vasoactive substances; (5) connective tissue production; and (6) regulation of SMC proliferation (71). On the other hand, SMC have only two major *in vivo* roles: contraction and production of vascular connective tissue proteins (72,73). EC maintain their differentiated phenotype in cell culture, while SMC lose contractility and enter an irreversible "proliferative state," retaining only the ability to divide and produce connective tissue (72). This section deals primarily with the effects of radiation on differentiated properties of cultured large vessel EC.

In vitro studies of radiation effects on EC function are summarized in Table 2. Most reports concern very early responses, occurring within three days after irradiation. Angiotensin-converting enzyme (ACE) is an EC-specific cell membrane-associated protein which converts AI to AII and metabolizes bradykinin (82). ACE activity decreased by 50-70% per attached cell 24 hours after a dose of 500 cGy (74).

Prostacyclin (PGI_2) is the principal prostaglandin synthesized by vascular EC. Its activities include inhibition of platelet aggregation and relaxation of vascular smooth muscle. EC cultures showed enhanced production of PGI_2 within 24 hours after doses as low as 400-600 cGy (19,75-77,81). This enhancement was related to increased *de novo* synthesis resulting from release of arachidonic acid from cell membrane stores and activation of cyclo-oxygenase (75,76,81). In contrast to the early response, the remaining attached EC showed a time-dose dependent decrease in PGI_2 synthetic capacity 7-21 days after irradiation (75). Late decrease in PGI_2 synthesis was observed in rabbit aortas irradiated *in vivo* and assayed 1-6 months after irradiation (83,84). Decreased PGI_2 production was also reported in human umbilical artery rings exposed *in vitro* to 200 cGy (85) and in irradiated rat and guinea pig abdominal aortas (86). A long-term decrease in PGI_2 synthesis might contribute to the increased risk of thrombosis and MI following mediastinal irradiation (31-34).

Radiation affects other EC interactions with the hemostatic system. Release of endothelial plasminogen activator is a major mechanism of tissue fibrinolysis. Release of plasminogen

Table 2. Effects of radiation on specialized function in confluent endothelial cell cultures

	Animal	Tissue	Radiation	Dose (cGy)	Time	Findings	Reference
1.	Bovine	Aorta	Co-60 gamma	500	24 hr	Decreased angiotensin-converting enzyme	Rubin (74)
2.	Bovine	Aorta	250 KV x-rays	500-5000	6-48 hr	Synthesis & release of PGI_2 into medium	Eldor (75)
				200-1200	7-21 dy	Decreased capacity to synthesize PGI_2	
3.	Bovine	Pulmonary	Cs-137 gamma	400-2000	24 hr	Increased capacity to synthesize PGI_2	Hahn (19,76)
4.	Bovine	Aorta	Co-60 gamma	50-500	4-24 hr	Increased PGI_2 synthesis	Rubin (77)
5.	Bovin	Aorta	Co-60 gamma	1000	72 hr	Inhibition of plasminogen activator release	Ts'ao (78)
6.	Human	Umbilical vein	Cs-137 gamma	$\geq$2000	24 hr	Release of von Willebrand protein	Sporn (79)
7.	Bovine	Pulmonary Artery	Cs-137 gamma	1500-3000	24 hr	Increased albumin permeability & disruption of intercellular junctions	Friedman (80)
8.	Bovine	Pulmonary Artery	Co-60 gamma	>600	24 hr	Increased uptake of Na^+-dependent amino acid and increased PGI_2 synthesis in cells grown on plastic dishes	Friedman (81)
9.	Bovine	Pulmonary Artery	Cs-137 gamma	$\geq$500	6-24 hr	Decreased uptake of Na^+-dependent amino acid uptake in cells on coverslips	Kwock (49)
10.	Bovine	Aorta	140 KV x-rays	800		Reduced inhibition of tumor cell destruction of smooth muscle matrix	Heisel (15)

activator by aortic EC was significantly reduced 72 hours after exposure to 1000 cGy (78). On the other hand, increased release of Von Willebrand protein was noted 24 hours after irradiation of human EC cultures (79). This protein, also called Factor VIII antigen, mediates platelet attachment to the subendothelium and hemostatic plug formation (87). Interestingly, release of von Willibrand protein showed a high threshold dose of 2000 cGy. Pre-irradiation protein labelling showed that release occurred from pre-formed intracellular stores (Weibel-Palade bodies). Inhibition of plasminogen activator release and stimulation of Von Willebrand protein release could promote radiation-induced vascular thrombosis. Decreased release of endothelial plasminogen activator could contribute to late radiation fibrosis by impairing degradation of interstitial fibrin, which provides a substrate for fibrogenesis (17).

Albumin transport across monolayers of pulmonary artery EC in serum-free medium was increased 24 hours after 1500 or 3000 cGy but was unchanged after 600 cGy (80). Increased albumin permeability was associated with disruption of intercellular junctional complexes and alterations in actin filament organization. All of these reverted to normal after incubation of cells in medium containing 10% serum. Na^+-dependent amino acid transport is an active process involved in cell growth, differentiation, and repair (88). Uptake of aminoisobutyric acid (AIB), a Na^+-dependent amino acid, in confluent EC cultures was increased 24 hours after doses >600 cGy (81). These increases correlated with increases in PGI_2 synthesis. The same group previously reported a decrease in AIB uptake (49). In this study, cells were trypsinized and seeded onto coverslips one day before irradiation. Recent trypsinization, growth state, and strength of attachment to basement membrane matrix can all influence cell membrane functions. Differences in these factors may account for the different EC responses in the two studies.

EC secrete a factor which inhibits degradation by tumor cells of connective tissue matrix proteins (16). Artificial vessel wall cultures made by growing bovine aortic EC on preformed layers of rat SMC were used to study the effects of radiation (800 cGy) on tumor cell degradation of SMC matrix (15). Pre-irradiation of SMC or EC alone caused a modest increase in degradation, while irradiation of both cell types prior to co-cultivation or pre-irradiation of the co-culture caused much larger increases. These findings suggest that EC-SMC interactions which are radiosensitive contribute to resistance to degradation. They suggest that functional changes following vascular irradiation may predispose toward invasion and metastasis of tumor cells.

Early responses of non-specialized EC activities have been reported (74). Irradiated EC cultures showed a time-dose dependent increase in cell detachment and in LDH and superoxide dismutase activities per ug DNA in the remaining attached cells. Increased release of LDH into the culture medium occurred at doses as low as 50 cGy. None of these responses were observed in SMC cultures. Ultrastructural alterations in cultured EC include vacuolization and increased numbers of lysosomes (24-48 hours), cell enlargement (72 hours), and reductions in polysomes and endoplasmic reticulum (7 days) (89). These changes were reversible at low doses (300-500 cGy) and progressive at high doses (>1500 cGy).

A number of early EC responses (e.g., alterations in PGI_2 synthesis, ACE activity, amino acid transport, release of von Willebrand protein) involve cell membrane-related processes. Thus early EC damage may reflect a common radiation-induced membrane abnormality (90). Long-term endothelial responses were investigated <u>in vivo</u> following hemithoracic irradiation (2500 cGy) of rats (91-95). After a transient increase, arterial perfusion decreased steadily to 40% of control by 6 months. ACE and plasminogen activator activities decreased to 15% and 50% of control, respectively. Doses between 1000 and 3000 cGy decreased ACE activity and perfusion at 2-6 months by about 3-4% per 100 cGy and increased collagen accumulation by 2% per 100 cGy (45). Pulmonary PGI_2 decreased transiently after one day, but then increased steadily to over twice control by 6 months, opposite to the pattern described previously (early increase, late decrease). After whole lung irradiation, it is difficult to distinguish direct effects of radiation on EC from the contributions of other cell types and from the secondary influence of ischemia in response to hypoxia from generalized alveolar injury (96).

A study of multilayered rat aortic SMC cultures revealed little or no change in culture morphology or protein synthesis on days 1, 3, or 9 after radiation doses up to 10,000 cGy (97). We do not know if production of specific proteins was altered. But these results indicate a gross lack of sensitivity of the SMC protein-synthesizing apparatus to doses which abolish replication.

To summarize, multiple specialized endothelial properties are acutely and chronically altered by radiation. Some early alterations may relate to cell membrane damage. The greater effect of radiation on morphology and proliferation of EC as compared with SMC might reflect natural diversion of endothelial metabolic machinery to maintenance of its differentiated state. The limited functional roles of SMC may be maintained even after large doses of radiation. If radiation does inhibit SMC synthesis of specialized matrix components, the very long turnover time of connective tissue molecules might delay expression of damage.

Radiobiology of Small Blood Vessels. Clonogenic survival curves for cultured capillary endothelial cells have not been reported. However, a number of attempts to assess *in vivo* capillary radiosensitivity using other end-points have been reported (98-106) (reviewed by Law (17)). In these studies, capillary proliferation was stimulated by local application of chemical irritants to rat subcutaneous tissue (98-100,104,105); surgical trauma to the dog cornea (101-103); and skin grafting in pigs (106). The end-point was inhibition of vascularization as a function of radiation dose. Most of these studies showed Do values of 170-265 cGy; n values of 2-7; Dq of 180-340 cGy; and incremental doses (D_2-D_1) of 180-290 cGy. (D_2-D_1 is a measure of sublethal damage repair and equals Dq if damage is fully repaired.) These values are slightly higher than the *in vitro* data for large vessel EC by clonogenic assay (Do 100-160 cGy; Dq 65-230 cGy). Two studies demonstrated increased capillary growth if a period of several days elapsed between irradiation and stimulation of growth, suggesting repair of potentially lethal radiation damage (99-105).

Anomalously high Do values (900-1000 cGy) were obtained if the time between stimulation and assay was short (100,106). This may reflect inadequate time for expression of radiation-induced damage. Vascular sprouting and capillary loop elongation in the cornea continue without EC proliferation for four days after exposure to 8000 cGy (107). EC migration *in vitro* is not inhibited by a dose of 1500 cGy (62). Therefore, the *in vivo* results may be influenced by processes other than cell survival, some of which are radioresistant (62,107,108). Since stimulation of angiogenesis was required to observe the end-point, these studies might not reflect the normal slowly proliferating state of the endothelium. Nonetheless, there was no evidence that capillary EC are unusually radiosensitive compared with other cell types including large vessel cells.

Proliferation kinetics of irradiated EC and SMC were studied in small arteries of an exteriorized mouse mesentery segment (109,110). After 3 weeks, there was no significant increase in EC or SMC (3H)-thymidine labelling index (LI). Initial EC depletion was followed by partial recovery from 9-12 months, and subsequent late depletion. SMC depletion occurred slowly and progressively after 12-18 months. EC recovery at 9-12 months without increased LI could not be explained by local proliferation and was attributed to repopulation from outside of the irradiated field. Two other studies did show increased labelling of capillary EC after irradiation of rabbit heart and mouse bladder (111,112). To fully investigate the role of EC repopulation, careful studies delineating relationships between parameters of cell proliferation; cell death; and vascular size, density, and cellularity are required.

Tumor vs Normal Tissue Endothelium. Recent data indicate large differences in proliferation kinetics of EC in human/animal tumors as compared with normal tissues (13,113). Single injection (3H)-thymidine LI's ranged from 10-20% in tumor compared with 0.01-1% in normal tissue endothelium. Corresponding values for continuous *in vivo* labelling for one week were 60-90% and 2-3%, respectively. These differences translated into a 20-2000 fold faster EC turnover rate in tumors compared with normal adult tissues, and have been proposed as a possible target for antitumor therapy (13).

These findings suggest that the acute effects of radiation should be more severe in tumor relative to normal tissue microvasculature. They may explain the clinical observation that bleeding from various types of tumors (e.g., cervix, endometrium, lung) can usually be controlled within a few days by relatively small doses of radiation (1000 cGy) (114). Whether this relates to functional endothelial injury resulting in platelet adhesion and clotting or from structural damage to vessels resulting in decreased blood flow is not known.

Solesvik and colleagues (12) followed vascular changes in human melanoma xenografts grown in nude mice after single radiation doses of 500-2500 cGy. Their findings suggested that at low doses (500 cGy) vascularization was increased at one week, due to angiogenesis or reopening of closed vessels due to reduced intratumor pressure. At moderate doses (1000-1500 cGy), the smallest vessels were severely damaged. At high doses (2000-2500 cGy), tumor shrinkage was greater than vascular loss, and tumors became supervascularized. Thus during a course of fractionated radiotherapy in the clinic, a balance between tumor cell killing and vascular response might influence radiocontrollability.

While many studies have demonstrated hypoxic cells in human and animal tumors (115-117), there is little clinical evidence to implicate radioresistant hypoxic cells as a cause of treatment failure. One such example is cervix carcinoma in which anemia is an independent poor prognostic feature and transfusion increases control rates (118,119). Increased tumor and stromal and vascularization correlates with radiocurability (120,121). Current approaches in tumors with significant hypoxic fractions include the use of agents which selectively kill or sensitize these cells to radiation (hypoxic sensitizers) (122); and the use of fluosol as a blood supplement capable of holding high concentrations of dissolved O_2 (123,124).

Molecular Biology of the Vessel Wall: Implications for Vascular Radiobiology. The recent explosion of knowledge of the molecular biology of growth factors (GF) and their receptors (R) may have major implications for understanding vascular pathobiology (125), including radiation injury. The relationships between GF-R interactions and normal/tumor cell proliferation, angiogenesis, wound repair, and oncogenesis are being investigated at fundamental levels. Polypeptide GF's which bind to specific high molecular weight R's are major mitogens for different classes of normal tissue cells: platelet-derived growth factor (PDGF)- mesenchymal cells (126); epidermal growth factor (EGF)- epithelial cells (127); and the endothelial cell growth factor/fibroblast factor (ECGF/FGF) family-endothelial cells (128,129). Some GF's and R's are encoded by cellular proto-oncogenes, the normal cell counterparts of viral transforming genes. PDGF is the product of c-sis which is related to the transforming gene of simian sarcoma virus (130); the EGF receptor is encoded by c-erb B, the homolog of the transforming gene of avian erythroblastosis virus (131).

Proliferating endothelial cells express c-sis mRNA and synthesize and release PDGF (132). However, they cannot respond to PDGF and are not transformed by simian sarcoma virus because they lack the PDGF receptor. PDGF released by EC can then stimulate proliferation of SMC, fibroblasts, and other mesenchymal cells. It also acts as a chemoattractant for these cells and stimulates connective tissue formation (133). PDGF is thought to play a role in atherogenesis by stimulating migration of SMC from media to intima and proliferation of intimal SMC (40). We might speculate that radiation accelerates atherogenesis by causing endothelial injury and release of PDGF. Fibrosis, a common generic late tissue effect of radiation may be mediated in part by release of "mitoattractants" for fibroblasts (such as PDGF and FGF) from endothelium and other sources. The desmoplastic reaction produced by many tumors may be due to release of mesenchymal cell GF's with resulting fibrogenesis.

Angiogenesis factors are an important group of substances which stimulate formation of new capillaries during normal embryonic development, in response to pathologic stimuli, and in tumors. A number of angiogenesis and endothelial cell growth factors bind to heparin, which is present on the EC surface. These factors can be purified by heparin-Sepharose chromatography (134-136). They include angiogenesis factors derived from a rat chondrosarcoma and a human hepatoma, and various endothelial cell mitogens (ECGF, acidic and basic forms of FGF, retinal-derived GF). An exciting new development is the isolation of angiogenin, a human tumor cell-derived angiogenesis factor (137-139). Angiogenin is a polypeptide containing 123 amino acids, whose three-dimensional structure and gene sequence are similar to pancreatic ribonuclease. It has ribonucleolytic activity, which may be a clue to angiogenin's physiologic mechanism.

Less is known about vascular growth inhibitors because their effects are more difficult to demonstrate. Angiogenesis inhibitors have been described and have been proposed as the basis of a new form of antitumor therapy (134,135). The endothelium produces a heparin-like inhibitor of SMC proliferation, which is active *in vitro* and *in vivo* (140). On the other hand, human arterial EC require heparin (plus other GF's) for optimal growth in cell culture (141). Another recent exciting discovery is that of a cell surface protein present in confluent contact-inhibited EC monolayers which inhibits migration and replication of actively growing EC (142). This protein may play a role in maintaining the integrity of the endothelial lining and preventing cell overgrowth. Could occlusion of arterioles by overproliferation of EC in irradiated tissues (9,17) be related to loss of this inhibitory protein? The paucity of information on the molecular biology of radiation injury should stimulate future investigations along these lines.

CONCLUSIONS

Radiation injury to small blood vessels (capillaries, arterioles, venules) contributes significantly to some forms of late normal tissue damage, such as radiation-induced hepatitis and dermal atrophy. In contrast to small vessels, large arteries and veins appear to be relatively radioresistant. Clinically, the most important consequence of irradiation of large arteries appears to be acceleration of atherogenesis. The difference in the radiation responses of small and large vessels cannot be explained in terms of the intrinsic radiosensitivities of isolated vascular cells. Regulation of smooth muscle cell growth in the normal and injured vessel wall is mediated by the endothelial cell through production of specific growth factors and inhibitors. A more thorough understanding of the pathogenesis of vascular radiation injury will require studies of the repopulation kinetics of endothelium, and of the influence of radiation on endothelial-smooth muscle cell interactions at the cell and molecular biologic levels.

ACKNOWLEDGEMENTS

This work was funded by NCI grant 5-PO1-CA-12662, by the Milton Fund 2016-2, and by a grant-in-aid from the American Heart Association, Massachusetts Affiliate, #13-538-856. We are grateful to Drs. Julian Down and C. Norman Coleman for critically reviewing this manuscript and to Anita Conti for help in preparation of the manuscript.

REFERENCES

1. Bloomer WD, Hellman S. Medical Intelligence. Current Concepts. Normal tissue responses to radiation therapy. N Engl J Med 293:80-83, 1975.
2. Fletcher GH. "Textbook of Radiotherapy", 3rd Edition (Lea and Febiger, Philadelphia), 1980, pp 181-184.
3. Withers HR, Peters LJ, Kogelnik HD. The pathobiology of late effects of irradiation. In: Radiation Biology in Cancer Research, (RE Meyn, HR Withers, eds, Raven Press, New York), 1980, pp 439-448.
4. Withers HR, Mason KA, Thames HD. Late radiation response of kidney assayed by tubule-cell survival. Brit J Radiol 59:587-595, 1986.
5. Michalowski A. The pathogenesis of late side-effects of radiotherapy. Clin Radiol 37:203-207, 1986.
6. Van der Kogel AJ. Mechanisms of late radiation injury in the spinal cord. In: Radiation Biology in Cancer Research, (RE Meyn, HR Withers, eds, Raven Press, New York), 1980, pp 461-470.
7. Fowler JF. Dose response curves for organ function or cell survival. Radiol 56:497-500, 1983.
8. Wheldon TE, Michalowski AS, Kirk J. The effect of irradiation on function in self-renewing normal tissues with differing proliferative organization. Brit J Radiol 55:759-766, 1982.

9. Hopewell JW. The importance of vascular damage in the development of late radiation effects in normal tissues. In: "Radiation Biology in Cancer Research," (RE Meyn, HR Withers, eds, Raven Press, New York), 1980, pp 449-459.
10. Casarett GW. "Radiation Histopathology," Volume I, CRC (CRC Press, Florida, Boca Raton), 1980, pp 39-50.
11. Sams A. Histological changes in the larger blood vessels of the hind limb of the mouse after X-irradiation. Int J Radiat Biol 9:165-74, 1964.
12. Solesvik OV, Rofstad EK, Brustad T. Vascular changes in a human malignant melanoma xenograft following single-dose irradiation. Radiat Res 98:115-128, 1984.
13. Denekamp J. Vascular endothelium as the vulnerable element in tumors. Acta Radiol (Oncol) 23:217-225, 1984.
14. Peters LJ, Mason KA, Withers HR. Effect of lung irradiation on metastases: radiobiological studies and clinical correlations, In: Radiation Biology in Cancer Research (RE Meyn, MR Withers, eds, Raven Press, New York), 1980, pp 461-470.
15. Heisel MA, Laug WE, Stowe SM, Jones PA. Effects of X-irradiation on artificial blood vessel wall degradation by invasive tumor cells. Cancer Res 44:2441-2445, 1984.
16. Heisel MA, Laug WE, Jones PA. Inhibition by bovine endothelial cells of degradation by HT 1080 fibrosarcoma, cells of extracellular matrix proteins. J Natl Cancer Inst 71:1183-1187, 1983.
17. Law MP. Radiation-induced vascular injury and its relation to late effects in normal tisues. Adv Radiat Biol 9:37-73, 1981.
18. Eassa EH, Casarett GW. Effect of epsilon-amino-n-caproic acid on radiation-induced increase in capillary permeability. Radiology 106:679-688, 1973.
19. Hahn GL, Menconi MJ, Polger P. The effect of gamma radiation on prostacyclin production in cultured pulmonary artery endothelium, In: Prostaglandins and Cancer (TJ Powles, RS Bockman, KV Honn, P Ramwell, eds, Alan R Liss, New York), pp 381-384, 1981.
20. Wharton JT, Delclos L, Gallager S, Smith JP. Radiation hepatitis induced by abdominal irradiation with the cobalt 60 moving strip technique. Am J Roentgenol 117:73-80, 1973.
21. Ingold JA, Reed GB, Kaplan HS, Bagshaw MA. Radiation hepatitis. Am J Roentgenol 93:200-208, 1965.
22. Hebard DW, Jackson KL, Christensen GM. Radiat Res 81:441-454, 1980. The chronological development of late radiation injury in the liver of the rat.
23. Hopewell JW, Wright EA. The nature of latent cerebral irradiation damage and its modification by hypertension. Brit J Radiol 43:161-167, 1970.
24. Ackerman LV. The pathology of radiation effect of normal and neoplastic tissue. Am J Roentgenol Radium Ther Nucl Med 114:447-459, 1972.
25. Rubin P, Keys H, Paulter C. Changing concepts in the tolerance of radioresistant and radiosensitive normal tissues and organs; In: Biological Bases and Clinical Implications of Tumor Radioresistance (GH Fletcher et al eds, Mason, New York), 1980, pp 175-195.
26. Moss WT, Brand WN, Battifora H. The heart and blood vessels, In: "Radiation Oncology, Rationale, Technique, Results" (CV Mosby Co, St. Louis), 1973, pp 248-256.

27. Asscher AW, Wilson C, Anson SG. Sensitization of blood vessels to hypertensive damage by X-irradiation. Lancet 1:580-583, 1961.
28. El-Naggar AM, El-Baz LM, Carsten AL, Chanana AD, Cronkite EP. Radiation-induced damage to blood vessels: a study of dose-effect relationship with time after X-irradiation. Int J Radiat Biol 34:359-366, 1978.
29. Lindsay S, Entenman C, Ellis EE, Geraci CL. Aortic arteriosclerosis in the dog after localized aortic irradiation with electrons. Circ Res 10:61-67, 1962.
30. Lindsay S, Kohn HI, Dakin RL, Jew J. Aortic arteriosclerosis in the dog after localized aortic x-irradiation. Circ Res 10:51-60, 1962.
31. Silverberg GD, Britt RH, Goffinet DR. Radiation-induced carotid artery disease. Cancer 41:130-137, 1978.
32. McReynolds RA, Gold GL, Roberts WC. Coronary heart disease after mediastinal irradiation for Hodgkin's disease. Am J Med 60:39-45, 1976.
33. Dollinger MR, Lavine DM, Foye LV. Myocardial infarction due to post-irradiation fibrosis of the coronary arteries. JAMA 195:316-319, 1966.
34. Cohn KE, Stewart JR, Fajardo LF. Hancock EW. Heart disease following radiation. Medicine 46:281-298, 1967.
35. Amromin GD, Gildenhorn HL, Solomon RD, Nadkarni BB, Jacobs ML. The synergism of x-irradiation and cholesterol-fat feeding on the development of coronary artery lesions. J Atheroscler Res 4:325-334, 1964.
36. Artom C, Lofland HB, Clarkson TB. Ionizing radiation, atherosclerosis, and lipid metabolism in pigeons. Radiat Res 26:165-177, 1965.
37. Fajardo LF and Lee A. Rupture of major vessels after radiation. Cancer 36:904-913, 1975.
38. Colquhoun J. Hypoplasia of the abdominal aorta following therapeutic irradiation in infancy. Radiology 86:454-456, 1966.
39. Fonkalsrud EW, Sanchez M, Zerubavel R, Mahoney A. Serial changes in arterial structure following radiation therapy. Surg Gynecol Obstet 145:395-400, 1977.
40. Ross R. The pathogenesis of atherosclerosis. N Engl J Med 314:488-500, 1986.
41. Goldberg ID, Botnick LE, Stemerman MB, Hannon EC. In vivo aortic smooth muscle cell kinetics: response to irradiation in the rat. Cell Tissue Kin 15:675, 1982.
42. Goldberg ID, Stemerman MB, Schnipper LE, Ransil BJ, Crooks GW, Fuhro RL. Vascular smooth muscle cell kinetics: a new assay for studying patterns of cellular proliferation in vivo. Science 205:920-922, 1979.
43. Rosen EM, Goldberg ID, Myrick KV and Levenson SE. Radiation survival properties of cultured vascular smooth muscle cells. Radiat Res 100:182-191, 1984.
44. Rosen EM, Goldberg ID, Myrick KV, Levenson SE. Radiation survival of vascular smooth muscle cells as a function of age. Int J Radiat Biol 48:71-79, 1985.
45. Rosen EM, Goldberg ID, Myrick KV, Halpin PA, Levenson SE. Survival and repair of potentially lethal radiation damage in confluent vascular smooth muscle cell cultures. Radiother Oncol 5:159-163, 1986.

46. Martin DF, Fischer JJ. Radiation sensitivity of cultured rabbit aortic endothelial cells. Int J Radiat Oncol Biol Phys 10:1903-1906, 1984.
47. Kwock L, Chaney EL, Guilford WB, Clark RL, Zaramba J. Effect of ionizing radiation on the pulmonary vasculature and isolated pulmonary artery endothelium. Int J Radiat Oncol Biol Phys 9 (suppl 1):166-167, 1983 (Abstr).
48. Nias AHW. The clinical significance of cell survival curves, In: "The Biological and Clinical Basis of Radiosensitivity" (M Friedman, ed, Charles C. Thomas, Springfield, IL), pp 156-169, 1974.
49. Kwock L, Doublas WH, Lin PS, Baur WE, Fanburg BL. Endothelial cell damage after gamma-irradiation in vitro: Impaired uptake of alpha-aminoisobutyric acid. Am Rev Respir Dis 125:95-99, 1982.
50. Rhee JG, Lee I, Song CW. The clonogenic response of bovine aortic endothelial cells in culture to radiation. Radiat Res 106:182-189, 1986.
51. Little JB. Repair of sublethal and potentially lethal radiation damage in plateau phase cultures of human cells. Nature 224:804-806, 1969.
52. DeGowin RL, Lewis LJ, Mason RE, Borke MK, Hoak JC. Radiation-induced inhibition of human endothelial cells replicating in culture. Radiat Res 68:244-250, 1976.
53. Griem ML, Fischer-Dzoga K, Dimitrievich GS. Radiosensitivity of vascular tissue. II. Differential radiosensitivity of endothelial and smooth muscle cells. Int J Radiat Oncol Biol Phys 9 (Suppl 1):167, 1983 (abstr).
54. Withers HR. The dose-survival relationship for irradiation of epithelial cells of mouse skin. Br J Radiol 40:187-194, 1967.
55. Withers HR, Elkind M. Microcolony survival assay for cells of mouse intestinal mucosa exposed to radiation. Int J Radiat Biol 17:261-267, 1970.
56. Thompson LH, Suit HD. Proliferation kinetics of x-irradiated mouse L cells studied with time lapse photography-II. Int J Radiat Biol 15:347-362, 1969.
57. Johnson LK, Longenecker JP, Fajardo LF. Differential radiation response of cultured endothelial cells and smooth myocytes. Anal Quant Cytol 4:188-198, 1982.
58. DeGowin RL, Lewis LJ, Hoak JC, Mueller AL, Gibson DP. Radiosensitivity of human endothelial cells in culture. J Lab Clin Med 84:42-48, 1974.
59. Rubin DB, Drab EA, Ward WF, Bauer KD. Cell cycle changes and cytotoxicity and irradiated cultures of bovine aortic endothelial cells. Radiat Res 108:206-214, 1986.
60. Fischer JJ. Proliferation of rat aortic endothelial cells following x-irradiation. Radiat Res 92:405-410, 1982.
61. Schwartz SM. Dynamic maintenance of the endothelium, In: "The Endothelial Cell - A Pluripotent Control Cell of the Vessel Wall," (DGS Thilo-Korner, RI Freshney, eds, S Karger, Basel), 1983, pp 113-125.
62. Sholley MM, Gimbrone MA, Cotran RS. Cellular migration and replication in endothelial regeneration: a study using irradiated endothelial cultures. Lab Invest 36:18-25, 1977.
63. Hirsch EZ, Chisolm GM, White HM. Reendothelialization and maintenance of endothelial integrity in longitudinal denuded tracks in the thoracic aorta of rats. Atherosclerosis 46:287-307, 1983.

64. Burkel WE, Vinter DW, Ford JW, Kahn RH, Graham LM, Stanley JC. Sequential studies of healing in endothelial seeded vascular protheses: Histologic and ultrastructure characteristics of graft incorporation. J Surg Res 30:305-324, 1981.
65. Hladovec J, Rossman P. Circulating endothelial cells isolated together with platelets and the experimental modification of their counts in rats. Thrombosis Res 3:665-674, 1973.
66. Bouvier CA, Gaynor E, Cintron JR, Bernhardt B, Spaet TH. Circulating endothelium as an indication of vascular injury. Thrombos Diathes Haemorrh Supple 40:163-168, 1970.
67. Fischer JJ, Unpublished data.
68. Narayan K, Cliff WJ. Morphology of irradiated microvasculature: A combined <u>in vivo</u> and electron microscopic study. Am J Pathol <u>106</u>:47-62, 1982.
69. Rosen EM, Mueller SN, Noveral JP, Levine EM. Proliferative characteristics of clonal endothelial cell strains. J Cell Physiol 107:123-137, 1981.
70. Repin VS, Dolgov VV, Zaikina OE, Novikov ID, Antonov AS, Nikolaeva MA, Smirnov VN. Heterogeneity of endothelium in human aorta. A quantitative analysis by scanning electron microscopy. Atherosclerosis 50:35-52, 1984.
71. Thilo-Korner DGS, Freshney RI, eds, "The Endothelial Cell - A Pluripotent Control Cell of the Vessel Wall," S. Karger, Basel, 1983.
72. Chamley-Campbell J, Campbell GR, Ross R. The smooth muscle cell in culture. Physiol Rev 59:1-61, 1979.
73. Ross R, Klebanoff SJ. The smooth muscle cell. I. <u>In Vivo</u> synthesis of connective tissue proteins. J Cell Biol <u>50</u>: 159-171, 1971.
74. Rubin DB, Drab EA, Ward WF, Smith LJ, Fowell SM. Enzymatic responses to radiation in cultured vascular endothelial and smooth muscle cells. Radiat Res 99:420-432, 1984.
75. Eldor A, Vlodavsky I, Hyam E, Atzmon R, Fuks Z. The effect of radiation on prostacyclin (PGI) production by cultured endothelial cells. Prostaglandins 25:263-279, 1983.
76. Hahn GL, Menconi MJ, Cahill M, Polgar P. The influence of gamma radiation on arachidonic acid release and prostacyclin synthesis. Prostaglandins 25:783-791, 1983.
77. Rubin DB, Drab EA, Ts'ao CH, Gardner D, Ward WF. Prostacyclin (PGI) synthesis in irradiated endothelial cells cultued from bovine aorta. J Appl Physiol 58:592-597, 1985.
78. Ts'ao C, Ward WF. Acute radiation effects on the content and release of plasminogen activator activity in cultured aortic endothelial cells. Radiat Res 101:394-401, 1985.
79. Sporn LA, Rubin P, Marder VJ, Wagner DD. Irradiation induced release of von Willebrand protein from endothelial cells in culture. Blood 64:567-570, 1984.
80. Friedman M, Ryan US, Davenport WC, Chaney EL, Strickland DL Kwock L. Reversible alterations in cultured pulmonary artery endothelial cell monolayer morphology and albumin permeability induced by ionizing radiation. J Cell Physiol 129:237-249, 1986.
81. Friedman M, Saunders DS, Madden MC, Chaney EL, Kwock L. The effects of ionizing radiation on the pulmonary endothelial cell uptake of alpha-aminoisobutyric acid and synthesis of prostacyclin. Radiat Res 106:171-181, 1986.

82. Dorer FE, Kahn JR, Lentz KE, Levine M, Skeggs LT. Hydrolysis of bradykinin by angiotensin-converting enzyme. Circ Res 34:824-827, 1974.
83. Sinzinger H, Firbas W, Cromwell M. Radiation induced alterations in rabbit aortic prostacyclin formation. Prostaglandins 24:323-329, 1982.
84. Sinzinger H, Cromwell M, Firbas W. Long-lasting depression of rabbit aortic prostacyclin formation by single-dose irradiation. Radiat Res 97:533-536, 1984.
85. Allen JB, Sagerman RH, Stuart MJ. Irradiation decreases vascular prostacyclin formation with no concomitant effect on platelet thromboxane production. Lancet 2:1193-1196, 1981.
86. Baluda VP, Sushkevich GN, Parshkov EM, Lukoyanova TI. Influence of gamma-rays Co-60 and fast neutrons on intravascular platelet aggregation and prostacyclin-like activity of the vascular wall. ZKb Med 62:243, 1982.
87. Sakariassen KS, Bolhuis PA, Sixma JJ. Human blood platelet adhesion to artery subendothelium is mediated by factor VIII - von Willebrand factor bound to the subendothelium. Nature 279:636-638, 1979.
88. Guidotti GG, Borghetti AF, Gazzola GC. The regulation of amino acid transport in animal cells. Biochim Biophys Acta
89. Lee S-L, Douglas WHJ, Lin P-S, Fanburg BL. Ultrastructural changes of bovine pulmonary artery endothelial cells irradiated in vitro with a 137-Cs source. Tissue Cell 15:193-204, 1983.
90. Koteles GJ. Radiation effects on cell membrane. Radiat Environ Biophys 21:1, 1982.
91. Ts'ao CH, Ward WF, Port CD. Radiation injury in rat lung. I. Prostacyclin (PGI) production, arterial perfusion, and ultrastructure. Rad Res 96:284-293, 1983.
92. Ward WF, Solliday NH, Molteni A, Port CD. Radiation injury in rat lung. II. Angiotensin-converting enzyme activity. Radiat Res 96:294-300, 1983.
93. Ts'ao CH, Ward WF, Port CD. Radiation injury in rat lung III. Plasminogen activator and fibrinolytic inhibitor activities. Radiat Res 96:301-308, 1983.
94. Ward WF, Molteni A, Ts'ao CH, Solliday NH. Radiation injury in rat lung. IV. Modification by D-Penicillamine. Radiat Res 98:397-406, 1984.
95. Ward WF, Molteni A, Solliday NH, Jones GE. The relationship between endothelial dysfunction and collagen accumulation in irradiated rat lung. Int J Radiat Oncol Biol Phys 11:1985-1990, 1985.
96. Down JD, Nicholas D, Steel GG. Lung damage after hemithoracic irradiation: Dependence on mouse strain. Radiother Oncol 6:43-50, 1986.
97. Martin BM. Unpublished data, 1986.
98. Reinhold HS, Buisman GH. Radiosensitivity of capillary endothelium. Brit J Radiol 46:54-57, 1973.
99. Reinhold HS, Buisman GH. Repair of radiation damage to capillary endothelium. Brit J Radiol 48:727-731, 1975.
100. Reinhold HS. Radiation and the microcirculation. In: "Frontiers of Radiation Therapy and Oncology," Vol 6 (JM Vaeth, ed, Karger, Basel), 1972, pp 21-43.
101. Gillette EL, Maurer GD, Severin GA. Endothelial repair of radiation damage following beta irradiation. Radiology 116:175-177, 1975.

102. Fike JR, Gillette EL. ^{60}Co gamma and negative pi meson irradiation of microvasculature. Int J Radiat Oncol Biol Phys 4:825-828, 1978.
103. Fike JR, Gillette EL, Clow DJ. Repair of sublethal radiation damage by capillaries. Int J Radiat Oncol Biol Phys 5:339-342, 1979.
104. Van den Brenk HAS. Macro-colony assay for measurement of reparative angiogenesis after x-irradiation. Int J Radiat Biol 21:607-611, 1972.
105. Van den Brenk HAS, Sharpington C, Orton C, Stone M. Effects of x-radiation on growth and function of the repair blasteoma (granulation tissue). II. Measurements of angiogenesis in the Selye pouch in the rat. Int J Radiat Biol 25:277-289, 1974.
106. Hopewell JW, Patterson TJS. The effects of previous x-irradiation on the revascularization of free skin grafts in the pig. Proc Brit Microcirc Soc, Northampton, 1971.
107. Sholley MM, Ferguson GP, Seibel HR, Montour JL, Wilson JD. Mechanisms of neovascularization: Vascular sprouting can occur without proliferation of endothelial cells. Lab Invest 51:624-634, 1984.
108. Yamaura H, Yamada K, Matsuzawa T. Radiation effect on the proliferating capillaries in rat transparent chambers. Int J Radiat Biol 30:179-187, 1976.
109. Hirst DG, Denekamp J, Travis EL. The response of mesenteric vessels to irradiation. Radiat Res 77:259-275, 1979.
110. Hirst DG, Denekamp J, Hobson B. Proliferation studies of the endothelial and smooth muscle cells of the mouse mesentery after irradiation. Cell Tissue Kinet 13:91-104, 1980.
111. Fajardo LF, Stewart JR. Capillary injury preceding radiation induced myocardial fibrosis. Radiology 101:429-433, 1971.
112. Stewart FA, Denekamp J, Hirst DG. Proliferation kinetics of the mouse bladder after irradiation. Cell Tissue Kinet 13:75-89, 1980.
113. Hobson B, Denekamp J. Endothelial proliferation in tumors and normal tissues. Continuous labelling studies, Brit J Cancer 49:405-413, 1984.
114. Halle JS, Rosenman JG, Varia MA, Fowler WC, Walton LA, Currie JL. 1000 cGy single dose palliation for advanced carcinoma of the cervix or endometrium. Int J Radiat Oncol Biol Phys 12:1947-1950, 1985.
115. Thomlinson RH, Gray LH. Histological structure of some human lung cancers and possible implications for radiotherapy. Brit J Cancer 9:539-549, 1955.
116. Evans HTS, Naylor PFD. The effect of oxygen breathing and radiotherapy upon the tissue oxygen tension of some human tumors. Brit J Radiol 36:418, 1963.
117. Denekamp J, Hirst DG, Stewart FA, Terry NHA. Is tumor radiosensitization by misonidazale a general phenomenon? Br J Cancer 41:1-9, 1980.
118. Bush RS. The significance of anemia in clinical radiation therapy (Editorial). Int J Radiat Oncol Biol Phys 12:2047-2050, 1986.
119. Hirst DG. Anemia: A problem or an opportunity in radiotherapy? Int J Radiat Oncol Biol Phys 12:2009-2017, 1986.

120. Siracka E, Siracky J, Pappova N, Revesz L: Vascularization and radiocurability in cancer of the uterine cervix. A retrospective study. Neoplasma 29:183-188, 1982.
121. Kolstad P. Intercapillary distance, oxygen tension, and local recurrence in cervix cancer. Scand J Clin Lab Invest 22 Suppl 106:145, 1968.
122. Brown MJ. Sensitizers and protectors in radiotherapy, Cancer 55:2222-2228, 1985.
123. Teicher BA, Rose CM. Perfluorochemical emulsions can increase tumor radiosensitivity. Science 223:934-936, 1984.
124. Fischer JJ, Rockwell S, Martin DF: Perfluorochemicals and hyperbaric oxygen in radiation therapy. Int J Radiat Oncol Biol Phys 12:95-102, 1986.
125. Schwartz SM. Molecular biology of the vascular wall. Proceedings of a meeting. Arteriosclerosis 4:647-656, 1984.
126. Stiles CD. The molecular biology of platelet-derived growth factor. Cell 33:653-655, 1983.
127. Carpenter G, Cohen S. Epidermal growth factor. Ann Rev Biochem 48:193-216, 1979.
128. Maciag T, Hoover GA, Weinstein R. High and low molecular weight forms of endothelial cell growth factor. J Biol Chem 257:5333-5336, 1982.
129. Schwartz SM, Gajdusek CM. Growth factors and the vessel wall. Prog Hemost Thromb 6:85-112, 1982.
130. Doolittle RF, Hunkapiller MW, Hood LE et al. Simian sarcoma virus oncogene, v-sis, is derived from the gene (or genes) encoding a platelet-derived growth factor. Science 221:275-277, 1983.
131. Stoschek CM, King LE. Functional and structural characteristics of EGF and its receptor and their relationship to transforming proteins. J Cell Biochem 31:135-152, 1986.
132. Collins T, Ginsburg D, Boss JM, Orkin SH, Pober JS. Cultured human endothelial cells express platelet-derived growth factor B chain: cDNA cloning and structural analysis. Nature 316:748-750, 1985.
133. Grotendorst GR, Seppa H, Kleinman HK, Martin GR. Attachment of smooth muscle cells to collagen and their migration toward platelet derived growth factor. Proc Natl Acad Sci USA 78:3669-3672, 1981.
134. Folkman J. How is blood vessel growth regulated in normal and neoplastic tissue? G.H.A. Clowes Memorial Award Lecture. Cancer Res 46:467-473, 1986.
135. Folkman J. Tumor angiogenesis. Adv Cancer Res 43:175-203, 1985.
136. Shing Y, Folkman J, Haudenschild C, Lund D, Crum R, Klagsbrun M. Angiogenesis is stimulated by a tumor-derived endothelial cell growth factor. J Cell Biochem 29:275-287, 1985.
137. Kurachi K, Davie EW, Strydom DJ, Riordan JF, Vallee BL. Sequence of cDNA and gene for angiogenin, a human angiogenesis factor. Biochemistry 24:5494-5499, 1985.
138. Palmer KA, Scheraga HA, Riordan JF, Vallee BL. A preliminary three-dimensional structure of angiogenin. Proc Natl Acad Sci USA 83:1965-1969, 1986.

139. Shapiro R, Riordan JF, Vallee BL. Characteristic ribonucleolytic activity of human angiogenin. Biochemistry 25:3527-3532, 1986.
140. Castellot JJ, Karnovsky MJ, Rosenberg RD. Endothelium, heparin, and regulation of vascular smooth muscle cell growth, In: "The Biology of Endothelial Cells" (EA Jaffe, ed, Martinus Nijhoff, Boston), 1984, pp 118-128.
141. Thornton SC, Mueller SN, Levine EM. Human endothelial cells: use of heparin in cloning and long-term serial cultivation. Science 222:623-625, 1983.
142. Heimark RL, Schwartz SM. The role of membrane-membrane interactions in the regulation of endothelial cell growth. J Cell Biol 100:1934-1940, 1985.

12

DERMAL DAMAGE FROM ULTRAVIOLET RADIATION

L.H. KLIGMAN
Department of Dermatology, The University of Pennsylvania School of Medicine. Philadelphia, PA 19104, U.S.A.

INTRODUCTION

Ultraviolet (UV) radiation is increasingly recognized as the cause of a vast number of changes in the skin of humans and animals. These include alterations at the molecular, cellular, tissue and systemic levels. In the recent past, much has been learned about the immediate effects in skin of acute UV exposure (i.e. sunburn) with its epidermal cell death, inflammation and vasodilation (1,2). With chronic exposure, many of the clinical and histologic effects can be seen only after decades. Visually, these are hyper-and hypopigmented macules, dry scaly, wrinkled skin with a variety of benign, pre-malignant and malignant neoplasms. All epidermal in origin, they lead, inexorably in humans, to the appearance we described as photo-aged (3). Underlying many of these visible manifestations are drastic changes in the dermis. These relate chiefly to destruction of mature collagen, with a compensatory overproduction of reticulin fibers, hyperplasia of elastic fibers eventuating in elastosis, increased levels of the glycosaminoglycans (GAGs) comprising the ground substance and changes in the microvasculature. First described in actinically damaged humans (4,5), systematic investigation required an animal model.

Early studies by Sams et al (6) with haired mice and severely damaging doses of UVB and UVC (30-50 minimal erythema doses or MEDs) reported a modest degree of elastic fiber hyperplasia. The UV dose was excessive and unrealistic in terms of human exposure. The development of the hairless mouse mutant made it possible, with far lower doses, to examine a whole range of photobiologic problems. Moreover, the UV-induced changes in the skin of this animal, ranging from acute responses (7) to those associated with chronic exposure such as tumorigenesis (8) and connective tissue damage (9) are comparable to those in human skin. The chronic effects of UVB, UVA and full spectrum solar radiation in these mice have been investigated in my laboratory.

P. Polgar (ed.), EICOSANOIDS AND RADIATION.

MATERIALS AND METHODS

Animals and treatment groups

Albino Skh:hairless-1 female mice, 6-8 weeks old, were distributed in the irradiation groups to provide 6-8 mice for sacrifice at each time or dose point. There were age-matched unirradiated controls for histologic comparison. The following were the irradiation groups: UVB, UVA (315-400nm), UVA (340-400nm) and solar simulating radiation (SSR) which combined UVB and UVA.

Radiation sources and schedules

The UVB source was a bank of 9 Westinghouse FS-20 "sunlamps" emitting 280-370nm radiation with a peak at 313nm. Doses equivalent to 2,3 and 6 MEDs (1 MED = .03 -.04 Joules/cm^2) were given thrice weekly for 10-30 weeks in the various experiments (9,10,11).

Full waveband UVA (315-400nm) was delivered by a forced air cooled 5000 W compact arc xenon solar simulator equipped with a 2 mm Schott WG 345 filter. Average daily doses of 35 Joules/cm^2 were given thrice weekly yielding a total dose of ~3,000 Joules/cm^2 in 34 weeks. Details have been published (11).

A bank of 8 black lights emitting UVA of 340-400nm with a peak at ~365nm was used to deliver a total of ~13,000 Joules/cm^2 given in 135 Joules/cm^2 doses thrice weekly for 34 weeks. The radiation was filtered through a sheet of mylar plastic to eliminate the shorter wavelengths (11).

SSR was obtained from the xenon source by passing the beam through a 1 mm Schott WG 320 filter. The spectral irradiance was similar to terrestrial solar UV radiation. Each exposure supplied ~.07 Joules/cm^2 UVB and ~7.0 Joules/cm^2 UVA. These were given thrice weekly for 34 weeks yielding totals of ~5.6 Joules/cm^2 UVB and ~560 Joules/cm^2 UVA (315-400). Details have been published (11).

Histology

In the various studies, animals were sacrificed at 10,20,30 and 45 weeks or at the end of the irradiation period and 15 weeks post-irradiation. Strips of dorsal skin, 2 cm long, were processed for light microscopy. The histochemical stains were: H & E, Luna's aldehyde fuchsin for elastic fibers (12), van Gieson's for collagen, Gordon and Sweet's for reticulin, Mowry's colloidal iron for total GAGs, alcian blue at pH 2.5 and 1.0 for hyaluronic acid and sulfated GAGs respectively.

Electron microscopy

Specimens were fixed routinely in 2% paraformaldehyde in 0.1M cacodylate-HCl at pH 7.4, post-fixed in osmium tetroxide, dehydrated in a graded series of alcohol and embedded in Epon 812. Thin sections were stained with saturated uranyl acetate and lead citrate. In some cases, tannic acid was added to the staining procedure (13) to enhance visualization of elastic fibers.

RESULTS

UVB effects

Because of its unequivocal implication in erythema and carcinogenesis, UVB was the obvious candidate as the perpetrator of human dermal photodamage. Actual experimental proof awaited not only the appropriate animal model, but the means to eliminate short wavelength UVC, absent in terrestrial sunlight yet present in most UVB sources.

Although, the hairless mouse has little elastic tissue (Fig. 1), severe elastosis (Fig. 2) was produced with UVB at a dose of 2 MEDs per exposure given thrice weekly for 30 weeks (11). In this experiment, the FS-20 UVB source was filtered with cellulose triacetate (CTA) film to attenuate UVC. In earlier studies, with higher doses (9,10) and no filtering, the elastosis was more extensive. This is attributed to the UVB dose (3 and 6 MEDs) rather than the UVC contamination because in other studies where different groups of animals were irradiated with and without CTA film, the elastosis was similar

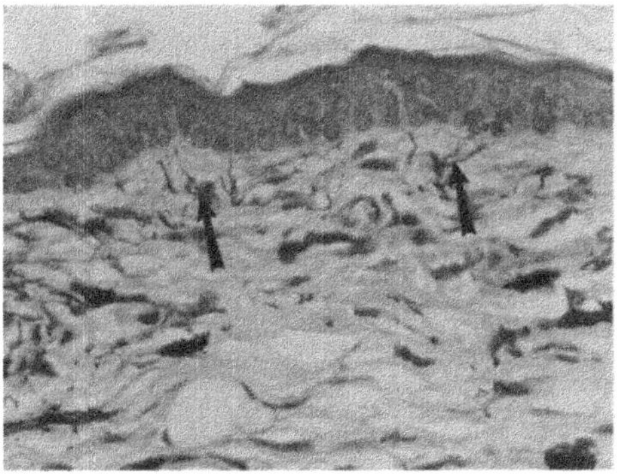

Fig. 1. Normal mouse skin. Elastic fibers (arrows) are fine and sparse.

Fig. 2. UVB-induced elastosis. Note large mast cells (arrows).

(unpublished results). Either insufficient UVC reaches fibroblasts in the dermis or there are no chromophores responding to this waveband that affect the elastic fiber biosynthetic pathway.

In UVB irradiated animals, fibroblasts were very numerous and large compared to controls. The dermis thickened with increased deposition of collagen, while a decreased affinity for van Gieson's stain suggested severe damage to pre-existing mature collagen. Reticulin fibers, normally limited to basement membrane zones, were prominent throughout the upper dermis (9) indicating new collagen synthesis. Finally, the GAGs comprising the ground substance were greatly increased. These findings were time and dose dependent. Biopsies at 10, 20 and 30 weeks showed progressively greater damage (10).

A notable response to UVB radiation was the induction of a severe dermal inflammatory infiltrate consisting of many lymphocytes and smaller numbers of eosinophils and neutrophils (11). Mast cells are conspicuous even in unirradiated hairless mouse skin where they mainly reside in the lower half of the dermis amidst the keratinizing cysts typical of this animal. UVB appears to recruit these cells into the upper collagen-containing dermis where they become very large and granular and are often in close association with the elastosis (9).

These histologic findings were confirmed by electron microscopy. The numerous fibroblasts had the characteristic ultrastructural features of high metabolic activity with greatly expanded cytoplasm and abundant widely dilated rough endoplasmic reticulum (9). Collagen bundles were in disarray and the increase in electron lucent ground substance was striking (Fig. 3).

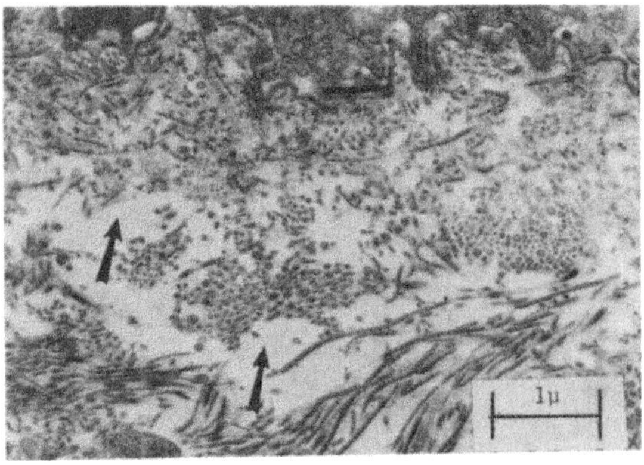

Fig. 3. Upper dermis after 10 weeks of UVB. Disrupted collagen bundles are surrounded by increased ground substance (arrows). Dermal-epidermal junction (—).

In addition to fraying and partial dissolution of large collagen fibers, there were numerous fibers of small diameters that are likely to be newly synthesized (Fig. 4a). Collagen fibers in unirradiated dermis have more uniform diameters (Fig. 4b). Elastic fibers, difficult to see with conventional stains but strikingly apparent with the addition of tannic acid, were clearly increased in number and often showed morphologic evidence of severe damage (14). The changes were reminiscent of those described in human photodamage by Braverman and Fonferko (15).

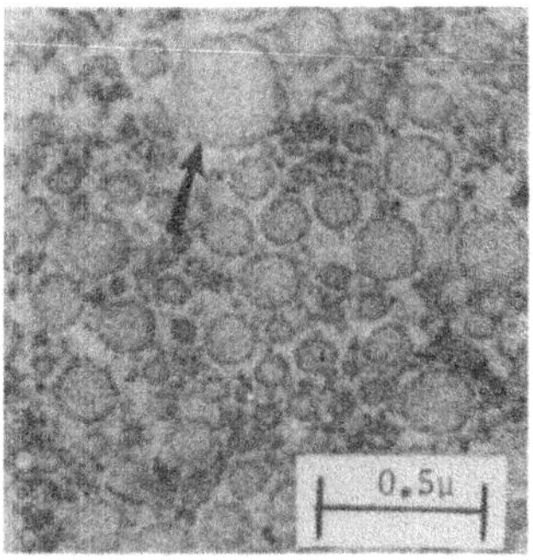

Fig. 4a. Collagen fibers in UVB-irradiated skin have widely divergent diameters in cross section and show evidence of dissolution (arrow).

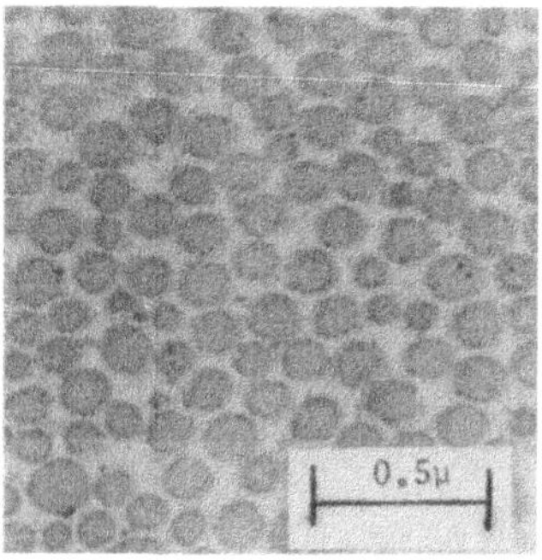

Fig. 4b. Collagen fiber diameters in unirradiated skin are more uniform.

Reprinted with permission from (ref. 9).

UVA effects

Recent work has shown that acute exposure to UVA (315-400nm) can, like UVB, produce erythema (2), endothelial cell enlargement, extravasation of blood cells and perivenular neutrophilic infiltrates in the dermis (16). However, the full extent to which UVA can induce dermal damage had not been studied. Because its effects on skin require doses that are ~1,000 times greater than UVB, the danger from UVA was thought to be negligible. However, it is 500-1,000 times more abundant in sunlight than UVB and, unlike UVB, it is present in large amounts all day and all year. Furthermore, its longer wavelengths allow more of it to reach dermal connective tissue and fibroblasts. Because of this, we were interested in examining the effects of UVA alone and in comparison to those of UVB (11).

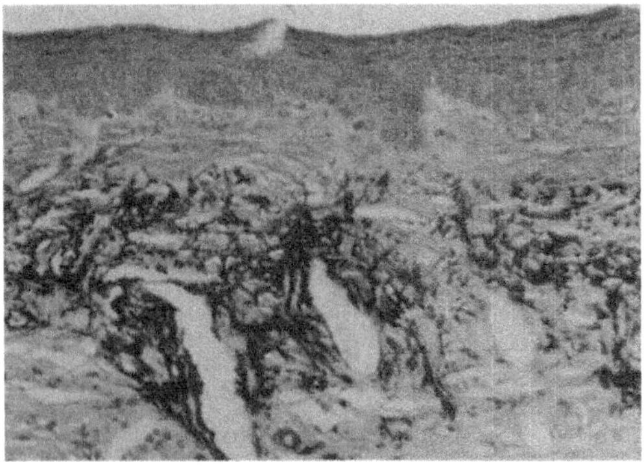

Fig. 5. Broad waveband UVA-induced elastosis extends deeply into the upper dermis. (Reprinted with permission from ref. 11).

Animals irradiated with a total of 3,000 Joules/cm^2 from the WG 345 filtered xenon source developed a significant amount of elastosis (Fig. 5). Less dense than that produced by a thousand-fold less UVB, it extended more deeply into the dermis. UVA from the black light source (340-400nm) produced only mild elastic fiber hyperplasia despite a total dose of ~13,000 Joules/cm^2. This strongly suggests that the shorter wavelengths (315-345nm), abundant in solar UVA but lacking or deficient in the black light spectrum, are those most responsible for UVA-induced photodamage.

Unlike UVB, UVA produced little or no morphologic or histochemical change in collagen. A possible explanation for this difference may be the finding that UVA does not evoke a severe inflammatory infiltrate in the dermis as does UVB (Fig. 6a and b), nor are mast cells recruited into the upper dermis.

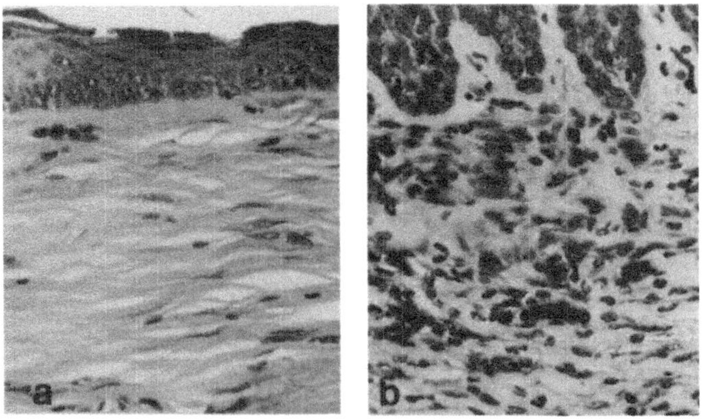

Fig. 6. a. UVA-irradiated skin with few cells, mainly fibroblasts. b. UVB-irradiated skin with a profuse inflammatory infiltrate. (Reprinted with permission from ref. 11).

This is in contrast to the neutrophilic infiltrate reported by Gilchrest et al in humans after acute UVA exposure (16). In studies with hairless mice and low dose, short term repeated UVB or UVA exposures, Johnston et al (17) reported no changes in collagen concentration as measured by hydroxyproline content. The course of irradiation was probably insufficient to produce a detectable difference. Interestingly, they did find a decrease in prolylhydroxylase activity in UVA irradiated animals which, if it persists, could lead to reduced collagen content and dermal atrophy. Histologically, we have not seen atrophy in our chronically irradiated animals.

Like UVB, UVA increased the GAG components of the ground substance. The distribution, however, was different. While UVB produced granular blue-staining material localized in the uppermost dermis, UVA-induced GAGs were deposited throughout the dermis, imparting to it a blue tone. Additionally, densely stained material was often deposited at the dermal-epidermal junction.

These findings provide compelling evidence that UVA radiation, especially that with a spectral quality similar to solar UV, is capable of inducing profound photodamage to dermal connective tissue in doses that are realistic in terms of human exposure.

Solar simulating radiation effects

The effects of UVA alone are interesting from a scientific viewpoint. However, in terrestrial solar radiation, it is inseparable from UVB, making the combined and possibly synergistic effects of the two broad wavebands of prime concern. These effects were studied by irradiating a group of mice with the xenon arc fitted with a WG 320 filter, providing SSR comprised of a hundred-fold more UVA than UVB (11).

In some respects the photodamage reflected the influence of both wavebands, while in others it was similar to UVB alone. Elastosis was severe and more dense than that produced by UVA alone. However, like with UVA alone, it extended deeply into the dermis. Collagen was damaged as with UVB and the inflammatory infiltrate, complete with mast cells, was present in the upper dermis. The increase in ground substance was similar to that produced by UVB while the darkly stained material at the dermal-epidermal junction was present as with UVA alone. Because of technical restraints, it was not possible to give these animals a UVA dose comparable to those receiving WG 345 filtered radiation. It is likely that the additional 2,500 Joules/cm^2 of UVA would have contributed even more strongly to the photodamage.

A surprising amount of connective tissue repair was noted 15 weeks after the UVB or SSR radiations were stopped (9,10,11). Prior to this observation, photodamage was thought to be irreversible. GAG levels gradually were reduced, but the most striking repair involved the collagen matrix. A band of new normal appearing collagen was deposited in the subepidermal dermis, pushing downward the old elastotic material (Fig. 7). The new dermis was free of any inflammatory infiltrate.

Fig. 7. Fifteen weeks post-UV. Sub-epidermal repair zone of new collagen with old elastosis at lower border (arrows). Fine new elastic fibers appear in the repair zone. Mast cells (*). (Reprinted with permission from ref. 9).

DISCUSSION

As a result of these studies, it has become clear that wavelengths within the UVB range are responsible for a large portion of photodamage to dermal connective tissue. UVA, long considered innocuous, has recently been shown to induce photodamage (2,11,16) in doses realistically attainable from solar radiation. It also adds to the photodamage induced by UVB. The shorter UVA wavelengths (315-340nm) appear to be the most destructive. It should be noted that frank elastosis was produced in another strain of hairless mice by Berger et al (18) with black lights which emit little or none of these wavelengths. It required, however, an inordinately high dose ($>$ 20,000 Joules/cm^2). In view of this, it is likely that all wavelengths of UV capable of reaching the dermis can produce some degree of photodamage if given in high enough doses.

The lack of collagen damage by UVA was interesting. While UV radiation may damage collagen directly via UV-labile pyridinoline cross-links (19) or other presently unknown mechanisms, most changes are probably indirect via enzymatic digestion, as suggested by Lovell (20). Such a possibility is supported by our findings, and those of Berger et al (18), that UVA neither damages collagen nor provokes inflammation, whereas with UVB radiation there is extensive inflammation and collagen destruction (9,11). It is well known that enzymes with specificities for dermal matrix can be found in cells of the inflammatory repertoire (21,22,23).

The role of UV radiation in the biosynthesis of mediators of inflammation, such as arachidonic acid metabolites, has been studied rather extensively in humans (24,25,26) and in animals (27,28). In all cases, these involved single UV exposures and mainly epidermal responses. Neither the effect of chronic UV irradiation nor the influence of these mediators on dermal connective tissue have been studied. Dermal effects are certainly implied by the temporal correlation, in humans, of peak levels of prostaglandins with the early stages of maximum erythema (24,25,26). However, this relates primarily to vascular damage. Ruzika et al (27) did examine arachidonic metabolism in the dermis of hairless mice acutely irradiated with UVC, UVB, UVA and UVA plus the photosensitizer 8-methoxy-psoralen (PUVA). They found a 5-fold increase in lipoxygenase activity, as measured by hydroxyeicosatetraenoic acid (HETE) production by tissue homogenates, only in PUVA treated animals. No cyclooxygenase activity was found in the dermis of normal mice and the effects of UV radiation, if any, were not reported. The lack of response may be a function of the short term UV exposures. Both UVB and UVA are apparently capable of stimulating the release of arachidonic acid from mouse fibroblasts in vitro (29,30). An increase in cyclooxygenase metabolites has also been reported (29). In addition, cyclooxygenase activity has been found in lymphocytes and eosinophils while both cyclooxygenase and lipoxygenase activities are reported in neutrophils, macrophages and mast cells (31), all components of chronically UVB irradiated hairless mouse dermis. Actinically damaged human skin is also chronically inflamed, a condition we call heliodermatitis (3).

Whether or not any of the arachidonic acid metabolites have any direct effect on dermal connective tissue remains to be investigated.

An indirect role in photodamage is likely where they probably serve as chemoattractants for the recruitment of the inflammatory infiltrate and its proteolytic enzymes.

REFERENCES

1. Gilchrest, B.A., Soter, N.A., Stoff, J.S. and Mihm, M.C. Jr. J. Amer. Acad. Dermatol. 5:411-422, 1981.
2. Kaidbey, K.H. and Kligman, A.M. J. Invest. Dermatol. 72:253-256, 1979.
3. Kligman, L.H. In: Dermatology Clinics. Aging Skin (Ed. B.A. Gilchrest), W.B. Saunders Co. Vol. 4, Philadelphia, 1986, pp.517-528.
4. Sams, W.M. Jr. and Smith, J.G. Jr. J. Invest. Dermatol. 37:447-452, 1961.
5. Kligman, A.M. J. Amer. Med. Assoc. 210:2377-2380, 1969.
6. Sams, W.M. Jr., Smith, J.G. Jr. and Burk, P.G. J. Invest. Dermatol. 43:467-471, 1964.
7. Cole, C.A., Davies, R.E., Forbes, P.D. and D'Aloisio, L.C. Photochem. Photobiol. 37:623-631, 1983.
8. Kligman, L.H. and Kligman, A.M. J. Nat. Cancer Inst. 67:1289-1297, 1981.
9. Kligman, L.H., Akin, F.J. and Kligman, A.M. J. Invest. Dermatol. 78:181-189, 1982.
10. Kligman, L.H., Akin, F.J. and Kligman, A.M. J. Invest. Dermatol. 81:98-102, 1983.
11. Kligman, L.H., Akin, F.J. and Kligman, A.M. J. Invest. Dermatol. 84:272-276, 1985.
12. Kligman, L.H. Amer. J. Dermatopath. 3:199-200, 1981.
13. Kajikawa, K., Yamaguchi, T., Katsuda, S. and Miwa, A. J. Electron Micro. 24:287-289, 1975.
14. Hirose, R. and Kligman, L.H. J. Invest. Dermatol. 86:481, 1986.
15. Braverman, I.M. and Fonferko, E. J. Invest. Dermatol. 78:434-443, 1982.
16. Gilchrest, B.A., Soter, N.A., Hawk, J.L.M., Barr, R.M., Black, A.K., Hensby, C.N., Mallet, A.I., Greaves, M.W. and Parrish, J.A. J. Amer. Acad. Dermatol. 9:213-219, 1983.
17. Johnston, K.J., Oikarinen, A.I., Lowe, N.J., Clark, J.G. and Uitto, J. J. Invest. Dermatol. 82:587-590, 1984.
18. Berger, H., Tsambaos, D. and Kaase, H. Z. Hautkr. 55:1510-1527, 1980.
19. Sakura, S., Fujimoto, D., Sakamoto, K., Mizuno, A. and Motegi, K. Can. J. Biochem. 60:525-529, 1982.
20. Lovell, W.W. Trans. St. Johns Hosp. Dermatol. Soc. 59:166-174, 1973.
21. Werb, Z., Banda, M.J. and Jones, P.A. J. Exp. Med. 152:1340-1357, 1980.
22. Mainardi, C.L., Hasty, D.L., Seyer, J.M. and Kang, A.H. J. Biol. Chem. 255:12006-12010, 1980.
23. Baugh, R.J. and Travis, J. Biochem. 15:836-841, 1976.
24. Camp. R.D., Greaves, M.W., Hensby, C.N., Plummer, N.A. and Warin, A.P. Br. J. Clin. Pharmacol. 6:145-148, 1978.
25. Black, A.K., Fincham, N., Greaves, M.W. and Hensby, C.N. Br. J. Clin. Pharmacol. 10:453-457, 1980.

26. Hawk, J.L.M., Black, A.K., Jaenicke, K.F., Barr, R.M., Soter, N.A., Mallett, A.I., Gilchrest, B.A., Hensby, C.N., Parrish, J.A. and Greaves, M.W. J. Invest. Dermatol. 80:496-499, 1983.
27. Ruzicka, T., Walters, J.F. and Printz, M.P. J. Invest. Dermatol. 81:300-303, 1983.
28. Burrall, B. and Ziboh, V.A. J. Invest. Dermatol. 86:643-648, 1986.
29. DeLeo, V.A., Hanson, D., Weinstein, I.B. and Harber, L.C. Photochem. Photobiol. 41:51-56, 1985.
30. DeLeo, V.A., Miller, S., Chew, S. and Scheide, S. Photochem. Photobiol. 43s:92, 1986.
31. Kragballe, K. and Voorhees, J.J. J. Invest. Dermatol. 81:293-296, 1983.

INDEX

GPSR Compliance

The European Union's (EU) General Product Safety Regulation (GPSR) is a set of rules that requires consumer products to be safe and our obligations to ensure this.

If you have any concerns about our products, you can contact us on ProductSafety@springernature.com

In case Publisher is established outside the EU, the EU authorized representative is:

Springer Nature Customer Service Center GmbH
Europaplatz 3
69115 Heidelberg, Germany

Batch number: 10755292

Printed by Printforce, the Netherlands